SpringerBriefs in Materials

W0079274

Series Editors

Sujata K. Bhatia, University of Delaware, Newark, DE, USA

Alain Diebold, Schenectady, NY, USA

Juejun Hu, Department of Materials Science and Engineering, Massachusetts Institute of Technology, Cambridge, MA, USA

Kannan M. Krishnan, University of Washington, Seattle, WA, USA

Dario Narducci, Department of Materials Science, University of Milano Bicocca, Milano, Italy

Suprakas Sinha Ray , Centre for Nanostructures Materials, Council for Scientific and Industrial Research, Brummeria, Pretoria, South Africa

Gerhard Wilde, Altenberge, Nordrhein-Westfalen, Germany

The SpringerBriefs Series in Materials presents highly relevant, concise monographs on a wide range of topics covering fundamental advances and new applications in the field. Areas of interest include topical information on innovative, structural and functional materials and composites as well as fundamental principles, physical properties, materials theory and design.

Indexed in Scopus (2022).

SpringerBriefs present succinct summaries of cutting-edge research and practical applications across a wide spectrum of fields. Featuring compact volumes of 50 to 125 pages, the series covers a range of content from professional to academic. Typical topics might include

- A timely report of state-of-the art analytical techniques
- A bridge between new research results, as published in journal articles, and a contextual literature review
- A snapshot of a hot or emerging topic
- An in-depth case study or clinical example
- A presentation of core concepts that students must understand in order to make independent contributions

Briefs are characterized by fast, global electronic dissemination, standard publishing contracts, standardized manuscript preparation and formatting guidelines, and expedited production schedules.

Nazia Hassan Kera · Sreejarani Kesavan Pillai ·
Suprakas Sinha Ray

Inorganic Ultraviolet Filters in Sunscreen Products

Status, Trends, and Challenges

 Springer

Nazia Hassan Kera
Department of Chemical Sciences
University of Johannesburg
Johannesburg, South Africa

Centre for Nanostructures and Advanced
Materials, DSI-CSIR Nanotechnology
Innovation Centre
Council for Scientific and Industrial
Research
Pretoria, South Africa

Suprakas Sinha Ray ⓘ
Centre for Nanostructures and Advanced
Materials, DSI-CSIR Nanotechnology
Innovation Centre
Council for Scientific and Industrial
Research
Pretoria, South Africa

Department of Chemical Sciences
University of Johannesburg
Johannesburg, South Africa

Sreejarani Kesavan Pillai
Centre for Nanostructures and Advanced
Materials, DSI-CSIR Nanotechnology
Innovation Centre
Council for Scientific and Industrial
Research
Pretoria, South Africa

ISSN 2192-1091   ISSN 2192-1105 (electronic)
SpringerBriefs in Materials
ISBN 978-3-031-64113-8   ISBN 978-3-031-64114-5 (eBook)
https://doi.org/10.1007/978-3-031-64114-5

This Springer imprint is published by the registered company Springer Nature Switzerland AG
The registered company address is: Gewerbestrasse 11, 6330 Cham, Switzerland

If disposing of this product, please recycle the paper.

Preface

The well-established dangers of exposure to ultraviolet radiation (UVR) in sunlight to human health and wellness have led to the development and use of a plethora of sunscreen products containing organic and/or inorganic UV filters that offer photoprotection from UVR for the prevention of skin damage, disease, and cancer. However, there is increasing concern over the safety of organic UV filters as they are susceptible to photodegradation on UV irradiation, which reduces product efficacy and generates harmful species that are detrimental to skin health. As such, inorganic UV filters are increasingly favoured over organic UV filters for use in sunscreens due to their advantages, such as high photostability, capacity for broad-spectrum UV protection, and skin compatibility. As ZnO and TiO$_2$ are currently the only two inorganic UV filters approved by regulatory bodies, their use in sunscreens for photoprotection applications is expected to continue unabated. However, the potential of ZnO and TiO$_2$ in sunscreens to cause genotoxic and cytotoxic effects in humans and organisms in the environment due to their inherent high photoactivity and small particle sizes, especially the nanoforms, has raised concerns over their use in sunscreens.

Moreover, sunscreen additives are reported to be an environmental hazard due to their potential to cause damage to marine and other ecosystems, such as coral reefs. Some of these concerns have translated into more stringent regulations regarding the use of ZnO and TiO$_2$ in photoprotection products in the market. There is a need for an updated review that provides a comprehensive status of inorganic UV filters by focussing on all the different aspects related to their use in sunscreens. This book, therefore, aims to provide an overview of inorganic UV filters for photoprotection applications by covering on the current literature related to their efficacy for photoprotection against UVR, formulation into sunscreen products, potential adverse effects to the human body, and impact in the environment on their inevitable release there, all in relation to their physiochemical properties, which largely influences all these aspects. As sunscreen products are governed by regulatory bodies, the current regulations related to the use of inorganic UV filters will also be mentioned in detail. As a way forward, the different modification techniques employed for improving the properties of ZnO and TiO$_2$ and alternative inorganic materials researched for use in sunscreens from previous studies will also be discussed in detail.

Review articles in the literature on the topic of UV filters typically provide either a general overview of both inorganic and organic UV filters by briefly touching on the different aspects related to their properties and other aspects related to their use in sunscreens or focus on a particular aspect in detail, such as their safety or environmental impact, for example. Few articles in the literature provide an updated overview of inorganic UV filters, while others concentrate on specific aspects only, such as safety, effectiveness, and environmental impact. However, to our knowledge, there is no book on this topic yet. Below are the unique selling points of this book:

- Provides a comprehensive overview of the recent development of inorganic UV filters used for photoprotection applications.
- The performance of the two inorganic UV filters approved for use in sunscreens, TiO$_2$ and ZnO, is assessed in relation to their physicochemical properties in terms of their efficacy towards UVR attenuation, formulation considerations such as product aesthetics and stability, safety aspects, and their potential risk to human health, and fate and effects in the environment.
- The regulations applicable to the use of TiO$_2$ and ZnO in sunscreens at present are also covered.
- Furthermore, the different modification strategies employed for diminishing the undesirable properties of ZnO and TiO$_2$ are gauged.
- Finally, the different inorganic materials studied as alternatives to ZnO and TiO$_2$ are presented, and their potential for use as UV filters is weighed up.

This book is ideal for chemists, material scientists, researchers, engineers (chemical and biomedical), and undergraduate and postgraduate students who are interested in this exciting field of research. It will also help industrial researchers and R&D managers who want to bring advanced inorganic UV filter-based sunscreen products into the market.

The authors would like to thank the Department of Science and Innovation, the Council for Scientific and Industrial Research, and University of Johannesburg, South Africa, for financial support. We express our sincerest appreciation to all colleagues, postdoctoral fellows, and students for their valuable contributions as well as the reviewers for their critical evaluation of the proposal and manuscript. We also thank the authors and publishers for permission to reproduce their published works. Our special thanks goes to Zachary Evenson at Springer for his encouragement, cooperation, suggestions, and advice during various phases of preparation, organization, and production of this book.

Johannesburg, South Africa   Nazia Hassan Kera
Pretoria, South Africa   Sreejarani Kesavan Pillai
Pretoria, South Africa   Suprakas Sinha Ray

Acknowledgments The authors would like to acknowledge the Department of Science and Innovation (C6E0085), Council for Scientific and Industrial Research

(C1E0080), and the University of Johannesburg, South Africa, for their financial support.

Conflict of Interest The authors declare they have no actual or potential conflict of interests.

Brief Introduction

Sunlight plays a vital role in photosynthesis and the propagation of life on Earth. However, research has shown that overexposure to high-energy ultraviolet radiation (UVR) in the sunlight adversely affects human health. Furthermore, although on track for recovery, the deterioration of the ozone layer by greenhouse gases in the atmosphere, arising from industrial processes, has resulted in increased levels of UVR reaching the surface of the Earth. Increased awareness of the harmful effects of UVR on human health and the importance of using sunscreen to protect the skin has led to the development of different types of sunscreen products over time, and a plethora of sunscreen products are commercially available. This brief aims to provide a comprehensive overview of the recent development of inorganic UV filters used for photoprotection applications. The performance of the two inorganic UV filters approved for use in sunscreens, TiO_2 and ZnO, is assessed in relation to their physicochemical properties in terms of their efficacy towards UVR attenuation, formulation considerations such as product aesthetics and stability, safety aspects, and their potential risk to human health, and fate and effects in the environment. The regulations applicable to the use of TiO_2 and ZnO in sunscreens at present are also covered. Furthermore, the different modification strategies employed for diminishing the undesirable properties of ZnO and TiO_2 are gauged. As a way forward, different inorganic materials studied as alternatives to ZnO and TiO_2 are presented, and their potential for use as UV filters is weighed up.

Contents

Chapter 1
Introduction

Sunlight plays a vital role in photosynthesis and the propagation of life on Earth. However, research has shown that overexposure to high-energy ultraviolet radiation (UVR) in the sunlight adversely affects human health [1–6]. Furthermore, although on track for recovery, the deterioration of the ozone layer by greenhouse gases in the atmosphere, arising from industrial processes, has resulted in increased levels of UVR reaching the surface of the Earth [1, 2]. Increased awareness of the harmful effects of UVR on human health and the importance of using sunscreen to protect the skin has led to the development of different types of sunscreen products over time, and a plethora of sunscreen products are commercially available [1–7]. In 2022, the revenue generated from the global sunscreen market was estimated to be 9.79 billion US dollars [8]. However, concerns have been raised over the efficacy and safety of UV filters, the active components that typically absorb, reflect, and scatter UVR, in sunscreen products for the photoprotection of skin [1–6]. The most widely used UV filters are either organic (chemical) or inorganic (physical) in nature [1–6]. Sunscreen products containing chemical filters are typically composed of organic compounds that can penetrate the skin barrier due to their intrinsic properties, such as low molecular weights and lipophilic nature [1–6].

Furthermore, the absorption of UVR by the organic compounds in chemical filters occurs through different mechanisms. It may involve the release of reactive oxygen species or degradation byproducts, which could damage the components of the skin or cause skin irritation or allergies [1–6]. Inorganic UV filters offer advantages over organic ones by providing broad-spectrum UV protection and high photostability [1–6]. Since the 1980s, inorganic UV filters have been incorporated into commercial sunscreen products and marketed for photoprotection applications [3, 7]. Currently, microsized and nanosized ZnO and TiO_2 are among the most widely used inorganic UV filters. They are the only two inorganic UV filters approved by regulatory bodies worldwide [9, 10]. However, there are challenges limiting the use of these inorganic UV filters in sunscreens related to their stability in formulations, product aesthetics, the tentative status of their permitted use by regulation bodies, and concerns over their safety of use with regard to human health, fate and potential effects when released in the environment [1–6]. Review articles in the literature on the topic of UV filters typically provide either a general overview of both inorganic and organic UV filters by briefly touching on the different aspects related to their properties and other aspects related to their use in sunscreens [1, 4, 5, 9, 11–15], or focus on a particular aspect in detail, such as their safety or environmental impact, for example [6, 16–23]. There are few studies in the literature that provide an updated overview of inorganic UV filters [2, 24–26], while others concentrate on certain aspects only such as safety, effectiveness, and environmental impact [3, 27–31]. This book aims to provide a comprehensive overview of the recent development of inorganic UV filters used for photoprotection applications. The performance of the two inorganic UV filters approved for use in sunscreens, TiO_2 and ZnO, is assessed in relation to their physicochemical properties in terms of their efficacy towards UVR attenuation, formulation considerations such as product aesthetics and stability, safety aspects, and their potential risk to human health, and fate and effects in the environment. The regulations applicable to the use of TiO_2 and ZnO in sunscreens at present are also covered. Furthermore, the different modification strategies employed for diminishing the undesirable properties of ZnO and TiO_2 are gauged. As a way forward, different inorganic materials studied as alternatives to ZnO and TiO_2 are presented, and their potential for use as UV filters is weighed up.

References

1. Egambaram OP, Pillai SK et al (2020) Materials science challenges in skin UV protection: a review. Photochem Photobiol 96(4):779–797
2. Manaia EB, Kaminski RCK et al (2013) Inorganic UV filters. Braz J Pharm Sci 49(2):201–209
3. Smijs TG, Pavel S et al (2011) Titanium dioxide and zinc oxide NPs in sunscreens: focus on their safety and effectiveness. Nanotechnol Sci Appl 4:95–112
4. Serpone N (2021) Sunscreens and their usefulness: have we made any progress in the last two decades? Photochem Photobiol Sci 20(2):189–244
5. Serpone N, Dondi D et al (2007) Inorganic and organic UV filters: Their role and efficacy in sunscreens and suncare products. Inorganica Chim Acta 360(3):794–802
6. Paiva JP, Diniz RR et al (2020) Insights and controversies on sunscreen safety. Crit Rev Toxicol 50(8):707–723
7. Ma Y, Yoo J et al (2021) History of sunscreen: an updated view. J Cosmet Dermatol 20(4):1044–1049
8. Market revenue of the sun protection market worldwide from 2013 to 2026. https://www.statista.com/forecasts/812522/sun-care-market-value-global#:~:text=In%202021%2C%20the%20global%20sun,of%20around%20500%20million%20dollar. Accessed 22 Feb 2024
9. Jesus A, Augusto I et al (2022) Recent trends on UV filters. Appl Sci 12(23):12003
10. National Research Council (2022) Review of fate, exposure, and effects of sunscreens in aquatic environments and implications for sunscreen usage and human health. The National Academies Press
11. Jesus A, Sousa E et al (2022) UV filters: challenges and prospects. Pharmaceuticals 15(3):263
12. Jansen R, Osterwalder U et al (2013) Photoprotection: part II. Sunscreen: development, efficacy, and controversies. J Am Acad Dermatol 69(6):S100–S109

13. Geoffrey K, Mwangi AN et al (2019) Sunscreen products: Rationale for use, formulation development and regulatory considerations. Saudi Pharm J 27(7):1009–1018
14. Parwaiz S, Khan MM (2023) Recent developments in tuning the efficacy of different types of sunscreens. Bioprocess Biosyst Eng 46(12):1711–1727
15. Mancuso JB, Maruthi R et al (2017) Sunscreens: an update. Am J Clin Dermatol 18(5):643–650
16. Adler BL, DeLeo VA (2020) Sunscreen safety: a review of recent studies on humans and the environment. Curr Dermatol Rep 9:1–9
17. Lozano C, Givens J et al (2020) Bioaccumulation and toxicological effects of UV filters on marine species. In: Sunscreens in coastal ecosystems: occurrence, behavior, effect and risk. The handbook of environmental chemistry, vol 94. Springer, Berlin/Heidelberg, Germany, pp 85–130
18. Chatzigianni M, Pavlou P et al (2022) Environmental impacts due to the use of sunscreen products: a mini-review. Ecotoxicology 31(9):1331–1345
19. Ruszkiewicz JA, Pinkas A et al (2017) Neurotoxic effect of active ingredients in sunscreen products, a contemporary review. Toxicol Rep 4:245–259
20. Gilbert E, Pirot F et al (2013) Commonly used UV filter toxicity on biological functions: review of last decade studies. Int J Cosmet Sci 35(3):208–219
21. Faco HAL, Guillermo MJ et al (2022) Potential systemic toxicity of UV filters in sunscreen: a review. Int J Res Publ Rev 3(5):3176–3191
22. Ballestín SS, Bartolomé MJL (2023) Toxicity of different chemical components in sun cream filters and their impact on human health: a review. Appl Sci 13(2):712
23. Lebaron P (2022) UV filters and their impact on marine life: state of the science, data gaps, and next steps. J Eur Acad Dermatol Venereol 36(S6):22–28
24. Schneider SL, Lim HW (2019) A review of inorganic UV filters zinc oxide and titanium dioxide. Photodermatol Photoimmunol Photomed 35(6):442–446
25. Wang SQ, Tooley IR (2011) Photoprotection in the era of nanotechnology. Semin Cutan Med Surg 30(4):210–213
26. More BD (2007) Physical sunscreens: On the comeback trail. Indian J Dermatol Venereol Leprol 73(2):80–85
27. Schilling K, Bradford B et al (2010) Human safety review of "nano" titanium dioxide and zinc oxide. Photochem Photobiol Sci 9(4):495–509
28. Nohynek GJ, Dufour EK (2012) Nano-sized cosmetic formulations or solid nanoparticles in sunscreens: a risk to human health? Arch Toxicol 86(7):1063–1075
29. Dréno B, Alexis A et al (2019) Safety of titanium dioxide nanoparticles in cosmetics. J Eur Acad Dermatol Venereol 33(S7):34–46
30. Yuan S, Huang et al (2022) Environmental fate and toxicity of sunscreen-derived inorganic ultraviolet filters in aquatic environments: a review. Nanomaterials (Basel) 12(4):699
31. Freitas Neto LL, Espósito BP (2023) Toxicity of zinc oxide to scleractinian corals and zooxanthellae: a brief review. Quim Nova 46(3):266–272

Chapter 2
Effects of UV Radiation in Sunlight on Skin

Sunlight plays a crucial role in photosynthesis and is, therefore, essential for life on Earth [1]. Sunlight is primarily composed of visible light (38.9%), UV radiation (UVR) (6.8%) and infrared radiation (IR) (54.3%) [2–4]. UVR consists of three components corresponding to different wavelength ranges: UVC (200–280 nm), UVB (280–315 nm), and long-wave UVA (315–400 nm), which is often subdivided into UVAI (340–400 nm) and UVAII (315–340 nm) [1–3, 5]. The intensity of the UVR reaching the surface of Earth depends on the season, latitude and longitude position, and height above sea level. It is also affected by cloud cover, ozone levels, and air pollutants [6]. Even though UVR comprises a small proportion of sunlight, exposure to UVR from sunlight is important for the production of Vitamin D3, a micronutrient essential for calcium absorption, and for maintaining good bone health in humans and animals [1–3]. UVR exposure is also linked to increased expression of β-endorphins in skin cells, which stimulate a sense of well-being [1, 3, 7]. However, overexposure to UVR in sunlight has been linked to human health issues such as skin damage and diseases/disorders such as erythema (sunburn), skin ageing and pigmentation, and carcinogenesis leading to malignant melanomas, squamous cell carcinomas and basal cell carcinomas [1–3, 6, 8–10]. UVB radiation, or medium-wave UV, directly damages DNA in skin cells due to its higher energy and is responsible for causing most skin cancers [1–3, 6, 8]. Exposure to UVB radiation also leads to erythema (skin redness) and sunburn [1–3, 6, 8]. UVA radiation, or long-wave UV, can penetrate up to the dermis layer of the skin, causing skin damage such as premature skin ageing and wrinkles [1–3, 6, 8]. Exposure to UVA and UVB radiation can also produce free radicals and reactive oxygen species, which can indirectly alter DNA, causing cancer [1–3, 6, 8]. UVC radiation, or short-wave UV, is mostly absorbed by the ozone layer and gases in the atmosphere, and only small doses reach the surface of the planet [1–3, 6, 8]. There is also little chance of UVC radiation penetrating skin due to complete absorption by skin chromophores [2]. According to the World Health Organisation (WHO), the occurrence of skin cancer cases, including both melanoma and non-melanoma skin cancer types, has been steadily rising with each

N. H. Kera et al., *Inorganic Ultraviolet Filters in Sunscreen Products*, SpringerBriefs in Materials, https://doi.org/10.1007/978-3-031-64114-5_2

new decade [1, 2, 6, 11]. The WHO estimates that non-melanoma and melanoma skin cancer cases will increase even further as stratospheric ozone levels decline and more UVR reaches the Earth's surface [1, 2, 6, 11].

References

1. Serpone N (2021) Sunscreens and their usefulness: have we made any progress in the last two decades? Photochem Photobiol Sci 20(2):189–244
2. Egambaram OP, Pillai SK et al (2020) Materials science challenges in skin UV protection: a review. Photochem Photobiol 96(4):779–797
3. Paiva JP, Diniz RR et al (2020) Insights and controversies on sunscreen safety. Crit Rev Toxicol 50(8):707–723
4. Barolet D, Christiaens F et al (2016) Infrared and skin: friend or foe. J Photochem Photobiol B, Biol 155:78–85
5. Diffey BL, Tanner PR et al (2000) In vitro assessment of the broad-spectrum ultraviolet protection of sunscreen products. J Am Acad Dermatol 43(6):1024–1035
6. Manaia EB, Kaminski RCK et al (2013) Inorganic UV filters. Braz J Pharm Sci 49(2):201–209
7. Jesus A, Sousa E et al (2022) UV filters: challenges and prospects. Pharmaceuticals 15(3):263
8. Smijs TG, Pavel S et al (2011) Titanium dioxide and zinc oxide NPs in sunscreens: focus on their safety and effectiveness. Nanotechnol Sci Appl 4:95–112
9. Serpone N, Dondi D et al (2007) Inorganic and organic UV filters: their role and efficacy in sunscreens and suncare products. Inorganica Chim Acta 360(3):794–802
10. Lu Y, Liu M et al (2021) Hydrogel sunscreen based on yeast/gelatin demonstrates excellent UV-shielding and skin protection performance. Colloids Surf B Biointerfaces 205:111885
11. World Health Organization (WHO). Radiation: ultraviolet (UV) radiation and skin cancer. https://www.who.int/news-room/questions-and-answers/item/radiation-ultraviolet-(uv)-radiation-and-skin-cancer. Accessed 22 Feb 2024

Chapter 3
UV Filters in sunscreens for photoprotection

Due to the harmful effects of UVR exposure on human skin, there has been an increase in the production and use of sunscreen products for the photoprotection of skin [1–7]. The most important components of sunscreen products are UV filters, which reflect, scatter, or absorb UVR to protect skin from harmful effects [1–6]. Sunscreen creams, first developed in the 1940s, have progressively improved to different product forms with access to a significantly larger number of active ingredients (UV filters), giving great flexibility in their formulations and compositions [1–7]. Although they are the critical functional component of sunscreens, the efficacy and human/environmental safety of many of the current UV filters have been brought into question lately [1–6]. UV filters are typically inorganic (physical) or organic (chemical) in nature, depending on their properties and mechanism of action [1–6].

Organic UV Filters

Organic UV filters are organic molecules containing an aromatic core linked to a conjugated electron system and carbonyl group. They are commonly referred to as chemical filters since their mode of operation involves the absorption of UVR [1, 2, 4, 6, 8]. Organic UV filters can be further categorized as UVA, UVB, or broad-spectrum (UVA and UVB) based on the UV wavelength range that they absorb, which in turn is determined by their chemical structure [4]. Most of the UV filters approved in the European Union (EU) and USA are organic filters [9]. Some organic UV filters approved for use in sunscreen products include benzophenone and its derivatives, aminobenzoic acid derivatives, salicylic acid derivatives, cinnamic acid derivatives, dibenzoyl methane derivatives, benzylidene camphor derivatives, triazines, benzimidazole derivatives and benzotriazole derivatives [1, 2, 8]. The absorption of UVR radiation results in conformational changes to the molecules of organic UV filters, as depicted in Fig. 3.1a [1, 2, 6, 8]. To return to the ground state, organic molecules

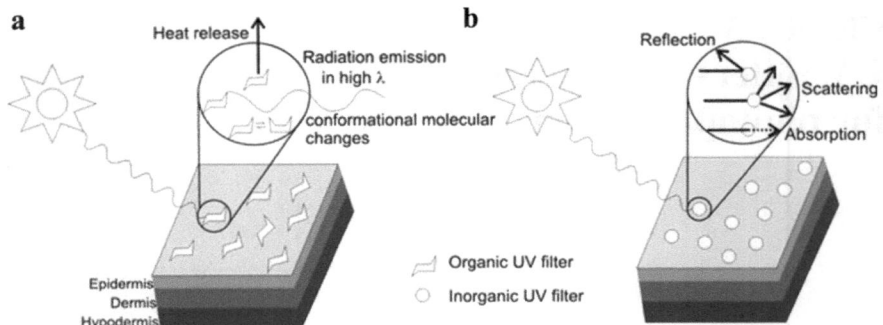

Fig. 3.1 Mode of operation of **a** organic and **b** inorganic UV filters. Reproduced with permission from [2], with slight modifications. Copyright 2013, Faculty of Pharmaceutical Sciences, University of São Paulo, Brazil

release energy as light or heat back to the skin through different mechanisms such as phosphorescence, fluorescence, or by degrading into byproducts [1, 2, 6, 8]. The major disadvantage of using organic UV filters in sunscreen products is their narrow UV absorption range and low photostability due to the photodegradation of molecules on absorbing UVR [1, 2, 6, 8].

The byproducts, free radicals, and other reactive species formed during the photodegradation of organic UV filters can damage skin cell DNA and skin components such as elastin and collagen and cause allergic/photoallergic contact dermatitis and sensitization [1, 2, 6]. Photodegradation of UV filters also lowers the efficacy of the sunscreen, thus compromising UVR protection efficacy [1, 2, 6]. As every organic UV filter has a specific absorption spectrum or wavelength range that effectively absorbs UVR, many organic UV filters cannot absorb both UVA and UVB radiations [1, 2, 6]. A combination of organic UV filters is often used to obtain sunscreen products with broad-spectrum UV protection [1, 2, 6]. However, the combination of organic UV filters in sunscreen products may potentially reduce their photostability [1, 6]. Human safety studies have also indicated the potential of organic UV filter molecules and photodegradation byproducts to penetrate the skin barrier, move into the dermis layer of skin, reach the blood, and subsequently cause toxic effects in the body [1, 6]. As such, some approved organic UV filters are no longer used in sunscreens such as US-approved para-aminobenzoic acid, trolamine, and cinoxate, and EU-approved, camphor benzalkonium methosulfate, benzophenone-5, and benzylidene camphor sulfonic acid [9]. The fate and effects of organic UV filters in the environment are also under scrutiny [10–15]. Commonly used organic filters have been detected in surface waters across the globe in numerous countries and the Arctic due to ineffectual wastewater treatment and major surface ocean currents [10, 14, 16]. Several studies have reported the harmful effects of organic UV filters released into aquatic systems on coral reefs and other marine life [10, 12–14, 17].

Inorganic UV Filters

Inorganic UV filters typically offer photoprotection of skin by absorbing, scattering, and reflecting UVR, as depicted in Fig. 3.1b [1, 2, 6]. Different inorganic materials investigated for use as UV filters include titanium dioxide, zinc oxide, iron oxides, talc, kaolin, mica, silica, calamine, red veterinary petrolatum, ichthammol, as well as carbonate and phosphate-based materials [1–4, 18]. Although all these materials can be used as UV filters, most offer poor UV protection, except for mineral oxides, namely titanium dioxide and zinc oxide.

Mineral Oxides

The most used inorganic filters in sunscreens in the market, which are the only two currently approved by the United States Food and Drug Administration (USFDA) and European Commission, are the mineral oxides, titanium dioxide (TiO_2), and zinc oxide (ZnO) [1–4, 7, 18]. Metal oxide-based inorganic UV filters are generally advantageous over organic UV filters due to their broad-spectrum coverage over the UVA/UVB region, higher photostability, lack of dermal penetration, and better skin compatibility due to a lower propensity for causing skin irritation, allergies, and sensitization, and as such are preferred by consumers with compromised skin [1–4]. Inorganic UV filters were previously frequently referred to as physical filters or sunblocks due to their mode of operation being attributed mainly to physical processes [19]. However, this is inaccurate as inorganic filters absorb UVR substantially, and the UV attenuation by inorganic UV filters is affected by both scattering and absorption [2, 5, 6, 20]. Inorganic UV filters such as TiO_2 and ZnO are semiconductors in nature. As such, their electronic structures consist of electronic bands, closely grouped atomic orbitals with similar energies, located in a valence band and a conducting band, the two of which are separated by a band gap, a region where no electronic states can reside, as shown in Fig. 3.2 [1, 3, 6]. It has been found that only under ~10% of the UV protection offered by inorganic filters is due to reflection and scattering, while the rest (over ~90%) is due to "semiconductor band gap mediated absorbance" of UVR [2, 6, 20].

The absorption of UVR of energy equal to or greater than the band gap energy leads to the excitation of an electron (e^-) from the valence band to the conduction band and results in the generation of a hole (h^+), an electron vacancy in the valence band that is highly localized [1, 3, 6]. The absorbed UVR is typically dissipated as heat or light [6]. However, the generated e^- and h^+ charge carriers can participate in oxidation–reduction reactions with acceptor or donor molecules on the surface of the crystal and from the surrounding media [1, 3, 6]. The UVR attenuation capabilities of metal oxide UV filters in sunscreens are influenced by their intrinsic properties such as refractive index, particle size, crystal phase, band gap energy, and surface area [1–5]. They are also affected by their dispersion within the formulation and

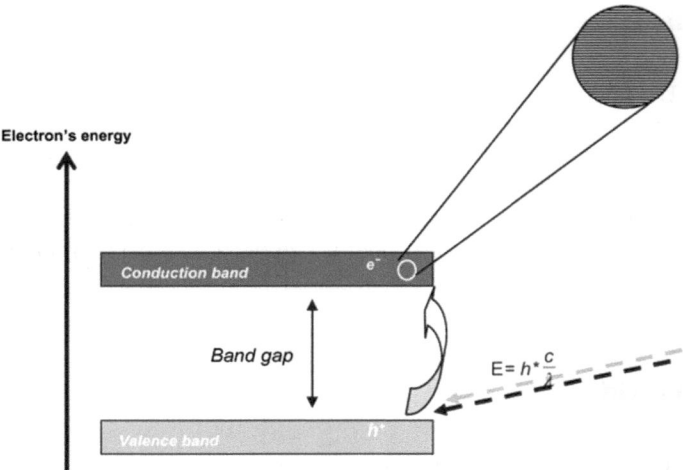

Fig. 3.2 Electronic structure of semiconducting materials and mechanism of UVR absorption. Reproduced with permission from [3]. Copyright 2011, Dove Medical Press Ltd.

nature of the medium [1–5]. The typical UVR absorption spectra of TiO_2 and ZnO UV filters are shown in Fig. 3.3 [2]. TiO_2 and ZnO UV filters with band gap energy values of 3.0–3.2 and 3.3 eV can typically absorb UVR of wavelengths shorter than 413–387 nm and 376 nm, respectively [4, 18].

Fig. 3.3 UVR absorbance spectra for TiO_2 and ZnO UV filters. Reproduced with permission from [2]. Copyright 2013, Faculty of Pharmaceutical Sciences, University of São Paulo, Brazil

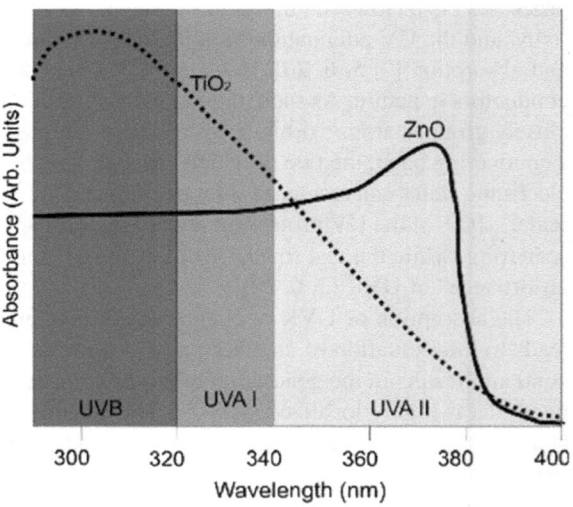

Titanium Dioxide

TiO_2, Titanium (IV) oxide, is a semiconducting metal oxide that occurs naturally, mainly in the mineral ilmenite [21]. Due to its white colour and opaque nature, TiO_2 is the most widely used white pigment globally and has found uses in plastics, coatings, ceramics, paper, paints and inks, and food colourants, among others [2, 21]. TiO_2 has three main crystalline forms, namely rutile, anatase, and brookite, with anatase being the most photoactive form [1–4, 6]. As brookite is rare, rutile and anatase are the forms of TiO_2 that are mostly used as UV filters in sunscreens [3]. The whiteness of TiO_2 can be attributed in part to its high refractive index, which results in its high reflectance [1–3]. Average refractive indices of 3.6 and 4.0 have been reported for anatase and rutile films, respectively [1–3]. TiO_2 is an indirect band gap semiconductor, with a valence band arising from the O 2p orbitals and the conduction band arising from the Ti 3d orbitals, that absorbs UV light mainly in the UVB region and also in the UVA II, as can be seen from its absorption spectrum in Fig. 3.3 [1–3].

Due to its broad UV absorption range, which extends from the UVA II to UVB region, TiO_2 is more frequently used in sunscreens than ZnO [2, 4, 22]. The band gap energy values reported for bulk phase rutile and anatase are $\sim$3.0 and $\sim$3.2 eV, corresponding to absorption edges of 413 nm and 387 nm, respectively [1, 3, 4]. Although anatase and rutile forms with tetragonal crystal structures are mainly used for photoprotection, the rutile form is the most thermodynamically stable phase, and its refractive index and density are higher than that of the anatase type [1–4]. Moreover, anatase is an aggressive free radical producer and adheres more strongly to the skin compared to rutile; hence, the use of the lesser photoactive rutile is favoured to produce safer sunscreen formulations [1–4].

Zinc Oxide

ZnO is found in nature in the rare mineral zincite [21, 23]. ZnO powder is used for many applications in industry, such as in paints, pigments, ceramics, cement, glass, rubber, lubricants, plastics, adhesives, fire retardants, and sealants, among others [23, 24]. ZnO is also widely used in electronic products such as light-emitting diodes, thin film transistors, liquid crystal displays, and semiconductors due to its exceptional optical, optoelectronic, ferromagnetic, and piezo-electric properties [24]. ZnO is also used in personal care products and cosmetics, as an astringent, as a bactericidal agent for treating various maladies and as a UV filter [2, 23, 24]. ZnO exists predominantly in one of two crystalline forms, the cubic-structured zinc blende and the hexagonal-structured wurtzite, which is the form that is more stable and thus more common [3, 23]. The refractive index values for ZnO, obtained from angle spectroscopic ellipsometry measurements in the 375–900 nm range, ranged between 2.0 and 2.3, below that of TiO_2 [1, 3]. ZnO is an n-type semiconducting material that has a direct and wide band gap and absorbs UVR mainly in the low-energy UVA region, and its absorption spectrum is shown in Fig. 3.3 [1, 3]. In particular, the wurtzite form of ZnO is a direct wide bandgap semiconductor with band gap energy ~3.4 eV with intrinsically high transparency over the whole visible range [3, 4]. Even though TiO_2 and ZnO have similar band gap energies, their UVR absorption properties differ, with ZnO absorbing mainly UVA radiation while TiO_2 absorbs more UVB radiation [1, 3]. This discrepancy has been partly attributed to the densely packed electronic states of TiO_2 that allow many absorption possibilities [1, 3].

In summary, UV filters in sunscreens are typically chemical (organic) or physical (inorganic) in nature and offer photoprotection by absorbing, reflecting, and scattering UVR. Even though the majority of approved UV filters are organic compounds, there are challenges associated with their use in sunscreens such as narrow UV absorption ranges, which necessitates the use of additives to extend UV coverage of sunscreens, and susceptibility to photodegradation, the byproducts of which can penetrate the skin and cause conditions such as allergic contact dermatitis. The low photostability and potential toxicity of some approved organic UV filters has even led to their current disuse in sunscreens. Inorganic UV filters are often favoured over organic UV filters due to their perceived higher photostability, better skin compatibility, and broader UV absorption ranges. However, the UV attenuation properties of inorganic UV filters are largely due to the inherently semiconducting nature of metal oxides, which has implications for sunscreen photostability, human safety and health, and their potential effects in the environment. The performance of sunscreens also depends on the properties of inorganic UV filters such as particle properties, photoactivity, refractive index, surface area, and other factors such as the properties and ingredients of the formulation medium. The use of inorganic filters in sunscreens is also dictated by current regulations. Therefore, these different factors must be taken into account in order to obtain a holistic view of inorganic UV filters, and these are discussed in detail in the sections that follow.

References

1. Egambaram OP, Pillai SK et al (2020) Materials science challenges in skin UV protection: a review. Photochem Photobiol 96(4):779–797
2. Manaia EB, Kaminski RCK et al (2013) Inorganic UV filters. Braz J Pharm Sci 49(2):201–209
3. Smijs TG, Pavel S et al (2011) Titanium dioxide and zinc oxide NPs in sunscreens: focus on their safety and effectiveness. Nanotechnol Sci Appl 4:95–112
4. Serpone N (2021) Sunscreens and their usefulness: have we made any progress in the last two decades? Photochem Photobiol Sci 20(2):189–244
5. Serpone N, Dondi D et al (2007) Inorganic and organic UV filters: Their role and efficacy in sunscreens and suncare products. Inorganica Chim Acta 360(3):794–802
6. Paiva JP, Diniz RR et al (2020) Insights and controversies on sunscreen safety. Crit Rev Toxicol 50(8):707–723
7. Ma Y, Yoo J et al (2021) History of sunscreen: an updated view. J Cosmet Dermatol 20(4):1044–1049
8. Jansen R, Osterwalder U et al (2013) Photoprotection: part II. Sunscreen: development, efficacy, and controversies. J Am Acad Dermatol 69(6):S100–S109

9. Pantelic MN, Wong N et al (2023) Ultraviolet filters in the United States and European Union: a review of safety and implications for the future of US sunscreens. J Am Acad Dermatol 88(3):632–646
10. Adler BL, DeLeo VA (2020) Sunscreen safety: a review of recent studies on humans and the environment. Curr Dermatol Rep 9:1–9
11. Lozano C, Givens J et al (2020) Bioaccumulation and toxicological effects of UV filters on marine species. In: Sunscreens in coastal ecosystems: occurrence, behavior, effect and risk, vol 94. The handbook of environmental chemistry. Springer, Berlin/Heidelberg, Germany, pp 85–130
12. Chatzigianni M, Pavlou P et al (2022) Environmental impacts due to the use of sunscreen products: a mini-review. Ecotoxicology 31(9):1331–1345
13. Mitchelmore CL, Burns EE et al (2021) A critical review of organic ultraviolet filter exposure, hazard, and risk to corals. Environ Toxicol Chem 40(4):967–988
14. Schneider SL, Lim HW (2019) Review of environmental effects of oxybenzone and other sunscreen active ingredients. J Am Acad Dermatol 80(1):266–271
15. Narla S, Lim HW (2020) Sunscreen: FDA regulation, and environmental and health impact. Photochem Photobiol Sci 19(1):66–70
16. Tsui MMP, Leung HW et al (2014) Occurrence, distribution and ecological risk assessment of multiple classes of UV filters in surface waters from different countries. Water Res 67:55–65
17. Miller IB, Pawlowski S et al (2021) Toxic effects of UV filters from sunscreens on coral reefs revisited: regulatory aspects for "reef safe" products. Environ Sci Eur 33:74
18. More BD (2007) Physical sunscreens: on the comeback trail. Indian J Dermatol Venereol Leprol 73(2):80–85
19. Geoffrey K, Mwangi AN et al (2019) Sunscreen products: rationale for use, formulation development and regulatory considerations. Saudi Pharm J 27(7):1009–1018
20. Cole C, Shyr T et al (2016) Metal oxide sunscreens protect skin by absorption, not by reflection or scattering. Photodermatol Photoimmunol Photomed 32(1):5–10
21. National Research Council (2022) Review of fate, exposure, and effects of sunscreens in aquatic environments and implications for sunscreen usage and human health. The National Academies Press
22. Jesus A, Augusto I et al (2022) Recent trends on UV filters. Appl Sci 12(23):12003
23. Raha S, Ahmaruzzaman M (2022) ZnO nanostructured materials and their potential applications: progress, challenges and perspectives. Nanoscale Adv 4(8):1868–1925
24. Ruszkiewicz JA, Pinkas A et al (2017) Neurotoxic effect of active ingredients in sunscreen products, a contemporary review. Toxicol Rep 4:245–259

Chapter 4
Characteristics and Performance of Inorganic UV Filter-Based Sunscreens

The most important characteristic of an ideal inorganic UV filter for photoprotection applications is its effective attenuation of UVA and UVB radiation through balanced processes of absorption, reflection, and scattering [1–4]. In addition, an inorganic UV filter should have an appropriate structure and properties such as high surface area, suitable refractive index, and high physical-, chemical-, and photostability [1–4]. An ideal UV filter should have low photoreactivity, be hypoallergenic and nontoxic, so as to not cause irritation and adverse effects when topically applied to the skin, and should not penetrate the skin layer and cause toxic effects in the body [1–5]. With regard to sunscreen formulations, an inorganic UV filter should be thin enough for easy formulation into products, impart good slippage, and appear transparent when applied on the skin, and be compatible with other components in the formulation [1–4]. In addition, inorganic UV filters that result in the environment at the end of life or as an unavoidable consequence of their use should not be harmful to ecosystems [6, 7]. The performance of commonly used inorganic UV filters is discussed below with respect to these desired characteristics.

Aesthetics and Stability of ZnO and TiO$_2$-Based Sunscreen Formulations

Older-generation sunscreens containing inorganic metal oxide UV filters had to be applied thickly on the skin in order to offer protection from UVR and had a few related drawbacks. These sunscreens appeared opaque on the skin and caused blackheads by blocking pores [8, 9]. They were prone to melting in the sun and caused stains on clothing [8, 9]. The undesirable white cast observed when sunscreens were applied on the skin was a result of the reflection and scattering of visible light by the relatively large-sized particles contained within the product having high refractive indices [5, 10–12]. Visible light is scattered to different extents by metal oxide particles depending on particle size and properties [13]. For instance, visible light is optimally scattered by ZnO and TiO$_2$ particles of 0.8 μm and 0.25 μm sizes, respectively [13, 14]. Below these sizes, visible light scattering decreases with particle size until sufficiently small sizes where visible light is then transmitted by the particles, which consequently appear transparent in thin films such as in sunscreens applied on the skin [10–14]. The size range of inorganic UV filter particles in sunscreens to appear transparent when applied is 10–20 nm for TiO$_2$ particles and 200 nm or below for ZnO particles, which is in relation to their refractive indices [11].

Obtaining stable sunscreen formulations is also a technical difficulty when incorporating large ZnO and TiO$_2$ particles in sunscreens, as these large particles can break emulsions and lead to unstable products with shorter shelf life [1, 9, 11]. As a result, in an attempt for manufacturers to obtain products with better aesthetic appeal and comfort in use that are easier to formulate, the sizes of ZnO and TiO$_2$ particles in sunscreens have been reduced significantly [5, 9–12]. As such, older-generation sunscreens have been largely superseded by sunscreens containing micronized, microfine or nanosized metal oxides [8, 9, 11]. However, it must be noted that there is no clear agreement on the general definition of the size range of micronized or microfine particles, and metal oxide particles of sizes spanning from nanometer to micron scale have been termed as such, with that of pigmentary grades ranging from 80 to 250 nm [8]. Furthermore, the transparency and efficacy of metal oxide particles in sunscreens are largely determined by particle size distribution rather than particle size [8]. Micronized particles of TiO$_2$ (150–300 nm) and ZnO (200–400 nm), both coated and uncoated, have widely been used in sunscreen products in the past few decades and are largely considered as effective for the photoprotection of the skin [11, 12, 15].

Pinnell et al. [16]showed that the microfine TiO$_2$ with a smaller average particle size range of 71 $\pm$ 38 nm (median 61 nm) appeared less transparent and whiter than microfine ZnO with a larger average particle size range of 117 $\pm$ 72 nm (median 94 nm) at varying film thicknesses and demonstrated this effect on Fitzpatrick type IV skin. This result was attributed to the lower refractive index and scatter efficiency of ZnO compared to that of TiO$_2$, which results in less visible light scattering by ZnO than the latter [16].

Efficacy of Sunscreen Formulations

Measurement of the Efficacy of a Sunscreen

The efficacy of a sunscreen product is typically based on the measurement of different factors which indicate UVA and UVB protection, mainly the UVA protection factor and sun protection factor, respectively, and to a lesser extent, the immune protection factor, which indicates protection against UV-induced immunosuppression [2, 4, 10, 17, 18].

The sun protection factor (SPF) value is the ratio of the minimal erythemal dose (MED), the amount of UVR required to cause erythema, for skin protected with topically applied sunscreen to that required to bring about the same response on unprotected skin with no sunscreen applied, as shown in Eq. 4.1 [2, 4, 10, 17, 18].

$$\text{SPF} = \frac{\text{MED for protected skin}}{\text{MED for unprotected skin}} \tag{4.1}$$

As such, SPF indicates the capacity of sunscreen to protect the skin from sunburn and prevent erythema caused by exposure to UVB and UVAII radiation [2, 4, 10, 17, 18]. The SPF does not indicate the capacity of a sunscreen for affording protection against UVAI radiation. The SPF value of a sunscreen can be determined through both *in vivo* and *in vitro* methods and can also be predicted from computer simulation models. The SPF determination of sunscreens, *in vivo*, typically involves the application of sunscreen at a dose of 2 mg/cm^2 to skin regions, preferably not exposed to sunlight or tanned for a minimum of 90 days, of consenting human volunteers. The selection of human subjects for *in vivo* SPF testing depends on several requirements, such as suitable skin type (skin types II, III, and IV of the Fitzpatrick scale) and skin condition (normal, unproblematic skin) for testing, among many others. Detailed guidelines for SPF determination in vivo can be obtained from internationally well-established methods, including the USFDA Final Rule 2011, ISO 24444:2019/AMD 1:2022, International SPF test method 2006/647/EC, and AS/NZS 2604 [2, 4, 17, 18].

As *in vivo* methods for the determination of sunscreen SPF are expensive and laborious, *in vitro* methods have also been developed in which tests are carried out on excised skin from human cadavers or suitable animal models or on synthetic substrates, with properties that mimic that of human skin [4, 10, 17, 18]. However, the controversy surrounding animal testing and its ban in some countries for the testing of cosmetics has led to the development and validation of *in vitro* methods for determining SPF based mainly on spectrophotometric analysis [4, 17]. These methods are typically less complicated, have lower costs, and have shorter analysis times than methods employing human and animal models [4, 17]. In these methods, the transmission or absorption of UVR passing through a solution or thin film of sunscreen on or in between quartz plates or other synthetic substrates such as polymethyl methacrylate (PMMA) plates is measured spectrophotometrically [2, 4, 10, 17, 18]. The SPF of the sunscreen, being equivalent to the inverse of the erythemic light transmitted by a film of sunscreen at each λ, can be obtained from the in vitro data by using Eq. 4.2, based on a model by Sayre et al. [19].

$$\text{SPF} = \frac{1}{T} = \frac{\sum_i \text{EE}(\lambda) \times I(\lambda)}{\sum_i \text{EE}(\lambda) \times I(\lambda) \times T(\lambda)} \tag{4.2}$$

where EE(λ) is the erythemal efficiency spectrum, $I(\lambda)$ is the solar simulator intensity spectrum, and $T(\lambda)$ is the spectroradiometer measure of sunscreen product transmittance related to absorbance (Abs) by Eq. 4.3 [19].

$$T(\lambda) = 10^{-\text{Abs}(\lambda)} \tag{4.3}$$

The EE $(\lambda) \times I(\lambda)$ values are constants at different wavelengths, as determined by Sayre et al. [19].

The SPF of sunscreens can also be calculated by the Mansur equation, as presented in Eq. 4.4, from the UV absorption spectra obtained in the 290–320 nm range from *in vitro* studies using diluted sunscreen formulation in liquid form [17, 20]. This simple equation, developed by Mansur et al. (1986), is based on substitution of the in vitro method proposed by Sayre et al. [19], as presented in Eq. 4.2.

$$\text{SPF} = \text{CF} \times \sum_{290}^{320} \text{EE}(\lambda) \times I(\lambda) \times \text{Abs}(\lambda) \tag{4.4}$$

where $\text{EE}(\lambda)$, $I(\lambda)$, $\text{Abs}(\lambda)$, and CF are the erythemal effect spectrum, solar intensity spectrum, spectrophotometric absorbance, and correction factor at wavelength λ, respectively [17, 20]. The values of the $\text{EE}(\lambda) \times I(\lambda)$ constants at wavelength λ, as determined by Sayre et al. [19], are also used for the computation of the SPF values via the Mansur equation [19, 20]. Based on their SPF values, the protection afforded by sunscreens against UVB radiation can be categorized as low (below SPF 15), medium (SPF 15–30), high (SPF 30–50), and very high (SPF > 50) [4]. The percentage of UVB radiation reaching the skin can also be obtained from a ratio of 100/SPF [17]. As an example, sunscreens of SPF 15, 20, 30, 40, and 50 protect the skin from 93.3, 95, 96.3, 97.8, and 98% of the UVB radiation where the formula to calculate sunscreen percentage absorption is $100 - (100/\text{SPF})$ [17]. However, the SPF values achieved under the conditions of real-life sunscreen use are typically lower than that indicated on the product as SPF testing is done under carefully controlled conditions and sunscreens are often applied incorrectly and at doses lower than 2 mg/cm^2, the specified dose for SPF testing [2, 4, 10, 17, 18]. The actual SPF attained is also dependent on a plethora of other factors, some of which are the thickness of the sunscreen layer, water contact, perspiration, dermal penetration properties, formulation medium properties, skin compatibility, and skin damage and diseases [2, 4, 10, 17, 18].

The UVA protection factor (UVA-PF) indicates the protection capacity of sunscreens against UVA radiation [2, 4, 10, 17, 18]. The UVA protection factor is the ratio of the minimal dose of UVR required to cause pigmentation (MPD) in skin protected with sunscreen to the MPD required to bring about the same response in unprotected skin with no sunscreen applied, as presented in Eq. 4.5 [4].

$$\text{UVA-PF} = \frac{\text{MPD}_p}{\text{MPD}_u} \tag{4.5}$$

where MPD_p and MPD_u are the minimal doses of UVR required to cause pigmentation in protected and unprotected skin, respectively [4]. The different methods employed to determine the UVA-PF of sunscreens, such as immediate pigment darkening (IPD) and persistent pigment darkening (PPD), are generally based on *in vivo* measurement of UVA-induced pigmentation on the skin [4, 10, 17, 18]. IPD is determined by measurement of the pigmentation that arises immediately on UVA irradiation on the skin and fades quickly. In contrast, the residual pigmentation, which generally persists for some time following exposure to UVA radiation, is measured to obtain the PPD [4, 10, 17, 18]. The PPD method for obtaining the UVA-PF has been adopted mainly in Japan [17, 18]. The PPD method does not provide an exact value of the UVA protection offered by a sunscreen but uses a rating to indicate the strength of UVA protection [17]. For example, a sunscreen with a PPD rating of 10 would theoretically protect skin from UVA-induced pigmentation for ten times longer than that of unprotected skin. However, the use of PPD ratings is still rare as a reproducible method for obtaining PPD scores is yet to be established, and as such, PPD values cannot be measured accurately.

For a sunscreen to be labelled as broad-spectrum, the USFDA and European Commission (EC) have recommended that the critical wavelength, which is the wavelength at which 90% of the area under the spectral absorbance curve falls, for the range 290–400 nm, should be 370 nm or longer [10, 17]. The EC has further recommended that the UVA-PF should be at least one-third of the overall SPF for a sunscreen product to be classified as broad spectrum. Three established standards, the Boot star method (United Kingdom), COLIPA method (EU), and FDA method (United States), have been widely adopted for the *in vitro* determination of the UVA-PF of sunscreens. In general, these methods involve UVR transmission measurement through a pre-irradiated sunscreen film, applied at a specified dose on a suitable substrate, with properties designed to mimic that of human skin, typically in the 290–400 nm region. The spectrophotometric data is then analysed to obtain the information relevant to the standard adopted, such as critical wavelength (FDA, COLIPA), ratio of absorbed UVA to absorbed UVB (Boots star rating system), percentage of incident UV radiation transmitted (Australian standard), and the ratio of UVA-PF to SPF (COLIPA standard).

A third factor, the immune protection factor (IPF), indicates the capacity of a sunscreen to prevent immunosuppression induced by UVR exposure, which is identified as a potential key event in the development of skin cancer [4, 21, 22–25]. Determination of the IPF typically involves contacting the skin with different allergens such as dichlorobenzene and nickel under UV-irradiation and measurement of the capacity of sunscreen for inhibition of either the sensitization or elicitation stages of immediate or delayed-type hypersensitivity reactions, respectively. However, the methods used usually require many volunteers and are too time-consuming and labour-intensive to become routine analysis methods [24]. Therefore, the use of IPF values as an additional indicator of the efficacy of sunscreens is at present hindered by the lack of standardized and less complicated methods for its determination [21, 24]. Even though the IPF values of sunscreens generally do not correlate with well their SPF values, their apparent better correlation to UVA-PF values suggests that higher UVA-PF values are required for obtaining good IPF values, further emphasizing the need for sunscreens with broad-spectrum coverage [22, 24–26].

Efficacy of ZnO and TiO₂-Based Sunscreen Formulations

The efficacy of metal oxide-based UV filters in sunscreens towards UVR attenuation depends mainly on their characteristics such as refractive index and particle properties (size, shape, dispersion, aggregation, and agglomeration) [2–4, 8]. It is also affected by the properties of the formulation medium and the thickness of the applied sunscreen film [2–4, 8]. As discussed previously, UVR is attenuated by inorganic UV filters through reflection, scattering, and absorption mechanisms, with TiO_2 mainly effective for absorbing UVR in the UVB region [10, 11]. In contrast, UVR absorption by ZnO occurs mainly in the UVA region [10, 11]. TiO_2 is typically reported to have a broader UV absorption range than ZnO; however, there are discrepancies in the UV absorption spectra data reported in the literature for ZnO and the different forms of TiO_2, rutile, and anatase [1, 2].

Bode et al. [27] investigated the efficacy of 8 of the 16 UV filters approved by the FDA that are most widely used in sunscreens in the US by using a solar-simulated light (SSL)-induced cutaneous squamous cell carcinoma (cSCC) mouse model. The efficacy of UV filters within a base lotion formulation was assessed individually or in combinations towards protection against damage due to SSL exposure and cSCC development, measured in terms of tumour volume and number. The results obtained showed that formulations containing ZnO (at 20%), individually, and ZnO (at 6.9%) in combination with an organic UV filter, octocrylene (at 7%), effectively attenuated both UVA and UVB radiation and were at least 99% effective in the reduction of tumour volume and cSCC development in hairless mice. In TiO_2-containing formulations, TiO_2 (12%), on its own, offered protection against UVB/shortwave UVA but was largely ineffective for preventing SSL-induced cSCC in hairless mice. In comparison, TiO_2 (at 6%) in combination with avobenzone (at 3%) and octocrylene (at 7%) was highly effective for UVA and UVB attenuation and for protecting against cSCC development. However, the inorganic UV filters used in this study were of pigment grades, which typically cause the undesirable white cast observed when sunscreens are applied to the skin, which necessitates reducing the inorganic filter concentration to overcome the undesired opaqueness and the addition of organic UV filters to maintain broad-spectrum UV protection [27].

There has been a move towards the use of micronized particles of TiO_2 and ZnO to achieve better dispersion of UV filters in sunscreen formulations and to reduce the reflection effects associated with larger particles [2, 3]. UVR absorption was found to vary with the particle size of metal oxide UV filters, with particle size on UVR absorption being more pronounced for TiO_2 [8]. Micronized ZnO particles were found to be effective for the absorption of UVR radiation across the UV region, covering wavelengths of 380 nm and under, while TiO_2 particles of sizes below 100 nm absorbed UVR of wavelengths 340–360 nm and below [8]. Information on the optical properties of different sizes of TiO_2 particles was investigated by application of Mie theory [1, 28]. As calculated from Mie theory, the UV/Visible attenuation curves showed that effective UV attenuation coupled with effective transmittance of visible light was obtained for particle sizes below 100 nm, with 50 nm found to be the optimal size [28]. Furthermore, Mie theory was used to determine the TiO_2 particle size ranges for effective attenuation of UVR at different wavelengths: 20–100 nm at $\lambda = 290$ nm (short-wave, UVB), 80–160 nm at $\lambda = 350$ nm (long-wave, UVA), and 120–180 nm at $\lambda = 400$ nm (long-wave, UVA) [1].

Mitchnick et al. [14] showed that uncoated microfine ZnO (Z-Cote), with an attenuation maximum of about 380 nm, was effective for the attenuation of UVA and UVB radiation and achieved transparent sunscreens with SPF 5.6 and SPF 18 by incorporating 5% and 15% Z-Cote, respectively, as the only active UV filter. The combination of Z-Cote (5%) and an organic UV filter, octyl methoxycinnamate (OMC) (7.5%), was found to be synergistic and resulted in an increased SPF of 21. Pinnell et al. [16] found that microfine ZnO showed lower transmittance and thus greater UVA attenuation than microfine TiO_2 for the UVR range 340–380 nm, at 2 and 6% concentrations by measuring diffuse reflectance *in vivo* on human skin.

However, the reduction of particle size of inorganic UV filters from the micro- to nanoscale leads to an increase in the band gap energies of both TiO_2 and ZnO, which shifts their corresponding UV–visible absorption spectra away from the UVA region and towards the UVB region, thus decreasing UVA absorption and compromising the broad-spectrum protection requirement as per USFDA recommendations [11, 12, 29]. One way of achieving the required UVA/B attenuation whilst retaining visible light transparency is to use a combination of ZnO and TiO_2 particles of appropriate micro- and nanosizes [2, 11]. In a related study, balanced UVA/B protection was achieved by a combination of ZnO, with particle sizes optimised for UVA absorption, and TiO_2, with particle sizes optimized for UVB protection, or through the combination of two different grades of ZnO particles, one of very small particle size at a high dose to afford a suitable SPF, and another of large particle size to provide UVA-PF [30]. Another method investigated to overcome the drawbacks of using NPs in sunscreen products is the use of ZnO and/or TiO_2 microparticles obtained through the controlled aggregation of NPs and microparticles [31, 32]. The UV absorption properties of TiO_2 (of anatase structure) and ZnO (of wurzite structure) microparticles and NPs and a ZnO–TiO_2 composite were compared in an investigation by Reinosa et al. [31]. The absorption curves showed that the UV absorption edge of TiO_2 NPs at 357 nm blue-shifted compared to that of TiO_2 microparticles at 391 nm. Correspondingly, the band gap energy of TiO_2 NPs (3.37 eV) was higher than that of TiO_2 microparticles (3.15 eV). In contrast, the same UV absorption edge (at 377 nm) and band gap energy of ~3.2 eV were observed for ZnO microparticles and NPs, showing an independence of particle size. The SPF curves of cream formulations containing different UV filter particles were also compared. The formulation containing TiO_2 NPs showed λ_{max} at ~319 nm and higher SPF values than formulations containing TiO_2 microparticles, which showed low SPF values and λ_{max} at ~360 nm. In contrast, formulations containing ZnO microparticles had higher SPF values than those containing ZnO NPs, with λ_{max} at ~368 nm in both cases. Based on these results, Reinosa et al. [31] prepared a ZnO–TiO_2 composite containing TiO_2 NPs (~15 wt%) dispersed onto ZnO microparticles (~85 wt%) by using two different preparation methods. The SPF and λ_c values of cream formulations containing the ZnO–TiO_2 composite (10 wt%) were 5.7 and 385.0 nm and 9.1 and 383.3 nm for the composite obtained from the standard and the dry dispersion methods, respectively. Raman spectroscopy results suggested that the dry dispersion method effectively anchored and dispersed TiO_2 NPs on ZnO microparticles, contributing to its better UV-absorption properties than that of the composite prepared from the standard method.

Another challenge associated with the use of NPs as UV filters is due to the tendency of primary particles to aggregate during synthesis and in formulations due to their relatively high interfacial energies [1, 11, 12, 33]. The formation of aggregates depends to a certain extent on factors such as ionic strength and pH of the background medium [11]. The tightly bound aggregates can further combine to form weakly bound agglomerates during heating and drying, as shown in Refs. [12, 33]. Commercially available sunscreen powders typically contain aggregates of sizes 30–150 nm of primary particles (sized 5–20 nm), which can further form agglomerates with sizes greater than 1 μm [12]. The aggregation and agglomeration of ZnO and TiO_2 NPs may reduce their UVR attenuation abilities since band gap energy is related to particle size. They may also decrease their apparent visible light transparency [1, 9, 11, 12, 33]. For example, TiO_2 NPs with varying average aggregate sizes (20, 50, and 100 nm) exhibited different UVA/B attenuation and visible light scattering properties [33]. TiO_2 NPs having an average aggregate size of about 100 nm were found to be effective for the attenuation of both UVA and UVB radiation but scattered visible light significantly. The 50 nm-sized particles showed greater UVB absorption at the expense of UVA protection but scattered visible light considerably less than the 100 nm-sized particles. The 20 nm-sized aggregates did not scatter visible light but had lower UVA and UVB protection than the larger sizes [33].

Reza Ghamarpoor et al. [34] investigated the effect of TiO_2 particle size on the performance of sunscreen formulations incorporating these UV filters in terms of SPF, UVA-PF, and critical wavelength measurements. The three commercially available TiO_2 samples used in this study included uncoated TiO_2, PVP polymer material-coated TiO_2, and polyisobutene-coated TiO_2. The particle sizes of the samples determined by photon correlation radiometry were higher than those indicated on the manufacturer labels, suggesting particle agglomeration had occurred. Three methods, including industrial homogenization, ball milling and ultrasonic homogenization, were then investigated to reduce the particle sizes of the three TiO_2 samples. The ultrasonic homogenization method resulted in the highest particle size reduction with particle sizes of 142.6, 254.8 and 262.8 nm obtained for the uncoated TiO_2, PVP polymer material-coated TiO_2, and polyisobutene-coated TiO_2 after 15 min of treatment, respectively. The sunscreen formulation containing the smaller-sized uncoated TiO_2 particles had a lower viscosity and slightly higher SPF and UVA-PF values than the formulations containing larger coated TiO_2 particles. The SPF values of all three formulations studied were found to decrease, with a corresponding increase in particle sizes, over a 1-month study period, which was attributed to particle agglomeration. For the uncoated TiO_2 sample, the SPF decrease and corresponding particle size increase over time were found to be highest for the formulation containing the lowest TiO_2 dosage. It was suggested that the effect of particle agglomeration was less marked in formulations containing higher TiO_2 dosages; however, the formulation stability was compromised at higher dosages. In the case of formulations containing the coated TiO_2 samples, the particle size increase and corresponding SPF decrease were independent of dosage. The SPF and UVA-PF values were found to decrease over a 1-month period to a greater extent at the higher formulation pH values studied of 6–7.5 than the lower pH values. As expected, the uncoated particles showed higher photocatalytic activity in methylene blue degradation studies than the coated particles. However, the formulations containing coated TiO_2 samples also met the requirements for offering broad-spectrum UV protection, which is advantageous due to the lower photocatalytic activity of coated TiO_2 samples.

The shape and morphology of NPs were also found to affect the UV attenuation properties of inorganic UV filters. In a study by Ilić et al. [35], the effect of TiO_2 nanowires, nanotubes and nanoplates on human keratinocytes (HaCaT skin cells) before and after exposure to UVB irradiation was investigated in vitro and assessed in terms of cell viability and mitochondrial membrane potential. The results showed that HaCaT skin cells were effectively protected from damage and death caused by UVB-irradiation by TiO_2 nanowires and nanoplates but not by nanotubes at the highest dose (300 mg/L) used in the study. Additionally, all three different TiO_2 NPs were not found to be cytotoxic at doses ranging from 10 to 300 mg/L.

Another widely used method for achieving broad-spectrum sunscreen products is by incorporating both inorganic and organic UV filters [2]. Lademann et al. [36] found that the SPF of organic UV filter-based sunscreens could be increased by the addition of inorganic ZnO and TiO_2 microparticles. This finding was explained as a synergistic effect involving the scattering of light photons by inorganic microparticles in the sunscreen, resulting in increased absorption by organic UV filters and a subsequent increase in SPF. Microparticles with high scattering properties increased the UV protection of organic sunscreens to a greater extent than those with lower reflection coefficients. This synergistic effect was also reported in a related study where the SPF values of six sunscreen formulations containing a combination of organic UV filters, with differing emulsifier systems, were significantly increased by the addition of aqueous TiO_2 dispersions [37].

However, a disadvantage of the combination of organic and inorganic UV filters is that the UV irradiation of ZnO and TiO_2 particles can bring about the degradation of organic UV filters or excipients in formulations, which will be discussed in a later section [37, 38].

NPs deposited within the skin, from topically applied sunscreen use, have the propensity to cause different physicochemical effects associated with particle–skin, particle–particle, and skin–particle–light interactions, which may alter their UV attenuation behaviour, as compared to that of NPs that remain on the skin surface [11]. Popov et al. [39] investigated the UV attenuation properties of TiO_2 particles embedded in the stratum corneum layer of skin. The tape-stripping technique, employed to determine the distribution of the spherical rutile TiO_2 particles (of 100 nm diameter) arising in the stratum corneum from topically applied sunscreen, showed that most of the particles occurred in the 0–3 μm depth of the stratum corneum from the surface. The researchers then employed Monte Carlo-based simulations to assess the interaction of TiO_2 particles embedded in the stratum corneum, of sizes ranging from 20 to 200 nm, with incident UV radiation of wavelengths of 310 and 400 nm. The scattering and absorption coefficients required as inputs for the simulation of a medium partially containing TiO_2 particles were calculated from Mie theory. They determined the particle sizes most effective for the attenuation of incident UV radiation of 310 and 400 nm to be 62 nm and 122 nm, respectively, for particles in the upper 1 μm depth of the 20-μm-thick stratum corneum. The simulations in this study were based on spherical TiO_2 particles, the properties of which are not characteristic of TiO_2 particles in commercial sunscreens, which are not always spherical and can be of other shapes, such as needle-like structures [40]. This study indicated that incident UV radiation can be effectively attenuated before reaching living cells by appropriately sized TiO_2 particles embedded in the stratum corneum. However, the relevance of this study is questionable as the UVR in sunlight covers a range of wavelengths, and the penetration of UV filters into the skin is generally altogether undesirable due to their potential to cause toxic effects there, as expanded upon in a later section, and as such sunscreen products should be designed to prevent dermal penetration of UV filter particles.

The public generally uses sunscreens to prevent sunburns and protect against skin cancer caused by UVR exposure [2, 41, 42]. However, even though commercial sunscreen products offering good technical performance are widely available in the market, the efficacy of sunscreen products in real-life scenarios depends largely on the implementation of recommended usage and application practices and correct sun protection behaviours [2, 41, 42]. This topic will not be expanded upon here but has been covered in several articles in the literature and is an important factor related to the use of sunscreens [2, 41, 42]. The cost of sunscreen products may also hinder their consistent and correct use [2].

Photostability of ZnO and TiO₂ in Sunscreens

Due to their photocatalytic nature, ZnO and TiO_2 have been widely studied as photocatalysts for various applications [1–3, 5, 10, 11]. However, this intrinsic property of ZnO and TiO_2 has raised concerns over their stability and the safety of their use in sunscreen products [1–3, 5, 10, 11]. Due to their photocatalytic ability, the absorption of UVR by ZnO and TiO_2 generates free radicals such as $^{\bullet}OH$ and reactive oxygen species (ROS) such as superoxide radicals ($O_2^{-\bullet}$), singlet oxygen ($^1O_2$) and H_2O_2, as illustrated simplistically in Fig. 4.1 and the photoreactivity of NPs is relatively much higher than larger forms of the metal oxides due to their smaller size and larger surface area [1–3, 5, 10, 11]. The absorption of UVR by a semiconducting metal oxide such as ZnO or TiO_2 results in the formation of conduction band electrons (e^-) and valence band holes (h^+). The scavenging of electrons by surface-adsorbed molecular oxygen generates superoxide radical anions ($O_2^{-\bullet}$), while the oxidation of hydroxyl groups (and chemisorbed H_2O) on the surface by holes forms hydroxyl radicals

Fig. 4.1 Scheme showing the generation of free radicals and ROS on the UV-irradiation of semiconducting metal oxides such as ZnO and TiO$_2$, in a simplistic fashion. Reproduced with permission from [3], with slight modifications. Copyright 2007, Elsevier Science Ltd.

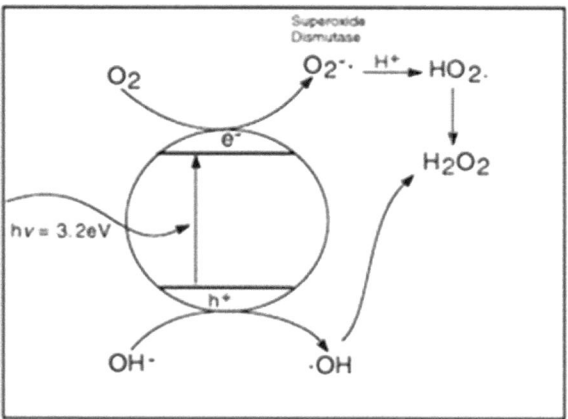

($^{•}$OH). At low pH, protonation of the superoxide radical anions occurs, generating hydroperoxyl radicals (HO$_2$$^{•}$) and then H$_2O_2$. The chain termination reaction of two $^{•}$OH anions can also form H$_2$O$_2$. The generated free radicals are highly oxidizing in nature and can participate in various reactions with different chemicals [1–3, 5, 10, 11]. Reactions on the surface or inside the skin can manifest as genotoxic and cytotoxic effects and skin damage [1–3, 5, 10, 11, 43, 44]. In sunscreen formulations, these reactions may affect the stability of inorganic UV filters or result in the degradation of organic UV filters and excipients, causing discolouring and adversely affecting product stability, efficacy, and shelf-life [1–3, 5, 10, 11].

Effect of the Photoreactivity of ZnO and TiO$_2$ on the Stability of Sunscreens

Even though TiO$_2$ and ZnO are typically of high photostability, these inorganic UV filters can cause the degradation of organic UV filters and other organic compounds in formulations, during storage or on the skin on UVR irradiation, due to their photocatalytic nature, thus reducing sunscreen efficacy and producing harmful organic byproducts [11, 38, 45]. As a consequence, the combination of a certain UV filter, avobenzone, with TiO$_2$ or ZnO in sunscreens is currently not approved by the USFDA, which has led to the more frequent use of oxybenzone in combination with these inorganic UV filters [46]. However, a study by Serpone et al. [47] found that oxybenzone was significantly photodegraded when present in sunscreens that also contain TiO$_2$. Oxybenzone, in the combination sunscreen, was degraded by 70% after 20 min of exposure to UVR, compared to that of oxybenzone on its own, which degraded by 50% after 260 min in aerobic aqueous media.

Ginzburg et al. [38] found that the addition of ZnO particles (6%) to organic UV filter-based sunscreens affected the photostability and efficacy of the product. Five sunscreen formulations with SPF 15 were prepared consisting of mixtures of common organic UV filters (avobenzone, octisalate, homosalate, octocrylene, oxybenzone, diethylamino hydroxybenzoyl hexyl benzoate, bisoctrizole) in different combinations, with compositions similar to products in the market. The composition and UV attenuation properties of the different as-prepared formulations and formulations with added uncoated microparticulate or nanoparticulate ZnO were analysed spectrophotometrically after exposure to UVR. The five as-prepared formulations containing only organic filters did not photodegrade after exposure to UVR. This was expected as the ratios of organic UV filters used in this study were similar to those of products in the market that have been optimized to prevent photodegradation. However, UVR irradiation of the five formulations with added ZnO resulted in photodegradation of the organic UV filters and a substantial decrease in UVA protection. The formulations containing ZnO were also found to increase morphological defects in zebrafish, when compared to the as-prepared formulations and formulations containing only ZnO, which suggested that the generated photodegradation byproducts were toxic [38]. In a related study, the yellowing of sunscreen formulations containing combinations of Butyl Methoxydibenzoylmethane (BMDM) with TiO_2 were investigated for illustrative purposes, as combination of these two UV filters is currently not permitted according to the FDA's rules related to sunscreen formulations [37]. The yellowing was more pronounced in formulations incorporating oil-based TiO_2 dispersions than aqueous TiO_2 dispersions, attributed to less interaction between the inorganic and organic UV filters due to separation of phases. The authors suggested the addition of a chelating agent or the use of silica-coated TiO_2 to prevent the formation of the coloured complexes which lead to the aesthetically undesirable yellowing of formulations.

In summary, large strides have been made towards the improvement of sunscreens incorporating metal oxide-based UV filters in terms of their aesthetic appeal and formulating ease and stability, especially when compared to the earlier generation sunscreens. This has largely been achieved through particle size reduction of TiO_2 and ZnO in sunscreens. However, the actual performance of sunscreen formulations towards photoprotection depends not only on metal-oxide particle size but also on the formulation properties and interactions arising thereof. The rigorous testing of sunscreen formulations is therefore essential, ideally in conditions mimicking real-life scenarios, and the determination of both SPF and UVA-PF via standardized methods to determine the spectrum of UV protection offered by a sunscreen product. Photostability studies are also necessary to determine the possible degradation of components and other effects arising in formulations due to the photocatalytic nature of ZnO and TiO_2. Studies in the literature often do not adequately report on the performance of metal oxide-based UV filters when contained in sunscreen formulations. As concerns have been raised over the use of ZnO and TiO_2 in sunscreens due to their properties, especially related to their particle size and photocatalytic nature, the safety of metal oxide-based sunscreens to human health and potential environmental impacts arising due to their use will be expanded upon in the sections to follow.

References

1. Manaia EB, Kaminski RCK et al (2013) Inorganic UV filters. Braz J Pharm Sci 49(2):201–209
2. Serpone N (2021) Sunscreens and their usefulness: have we made any progress in the last two decades? Photochem Photobiol Sci 20(2):189–244
3. Serpone N, Dondi D et al (2007) Inorganic and organic UV filters: Their role and efficacy in sunscreens and suncare products. Inorganica Chim Acta 360(3):794–802
4. Geoffrey K, Mwangi AN et al (2019) Sunscreen products: rationale for use, formulation development and regulatory considerations. Saudi Pharm J 27(7):1009–1018
5. Paiva JP, Diniz RR et al (2020) Insights and controversies on sunscreen safety. Crit Rev Toxicol 50(8):707–723
6. Lozano C, Givens J et al (2020) Bioaccumulation and toxicological effects of UV filters on marine species. In: Sunscreens in coastal ecosystems: occurrence, behavior, effect and risk. The handbook of environmental chemistry, vol 94. Springer, Berlin/Heidelberg, Germany, pp 85–130
7. Chatzigianni M, Pavlou P et al (2022) Environmental impacts due to the use of sunscreen products: a mini-review. Ecotoxicology 31(9):1331–1345
8. More BD (2007) Physical sunscreens: on the comeback trail. Indian J Dermatol Venereol Leprol 73(2):80–85
9. Lowe NJ (2006) An overview of ultraviolet radiation, sunscreens, and photo-induced dermatoses. Dermatol Clin 24(1):9–17
10. Egambaram OP, Pillai SK et al (2020) Materials science challenges in skin UV protection: a review. Photochem Photobiol 96(4):779–797
11. Smijs TG, Pavel S et al (2011) Titanium dioxide and zinc oxide NPs in sunscreens: focus on their safety and effectiveness. Nanotechnol Sci Appl 4:95–112
12. Nery ÉM, Martinez RM et al (2021) A short review of alternative ingredients and technologies of inorganic UV filters. J Cosmet Dermatol 20(4):1061–1065
13. Araujo TS, de Souza SO (2008) Protetores solares e os efeitos da radiação ultravioleta. Scientia Plena 4(11):114807
14. Mitchnick MA, Fairhurst D et al (1999) Microfine zinc oxide (Z-Cote) as a photostable UVA/UVB sunblock agent. J Am Acad Dermatol 40(1):85–90
15. Ma Y, Yoo J et al (2021) History of sunscreen: an updated view. J Cosmet Dermatol 20(4):1044–1049
16. Pinnell SR, Fairhurst D et al (2000) Microfine zinc oxide is a superior sunscreen ingredient to microfine titanium dioxide. Dermatol Surg 26(4):309–314
17. Parwaiz S, Khan MM (2023) Recent developments in tuning the efficacy of different types of sunscreens. Bioprocess Biosyst Eng 46(12):1711–1727
18. Young AR, Claveau J et al (2017) Ultraviolet radiation and the skin: photobiology and sunscreen photoprotection. J Am Acad Dermatol 76(3S1):S100–S109
19. Sayre RM, Agin PP et al (1979) A comparison of in vivo and in vitro testing of sunscreening formulas. Photochem Photobiol 29(3):559–566
20. Dutra EA, Da Costa E, Oliveira DAG et al (2004) Determination of sun protection factor (SPF) of sunscreens by ultraviolet spectrophotometry. Braz J Pharm Sci 40(3):381–385
21. Jansen R, Osterwalder U et al (2013) Photoprotection: part II. Sunscreen: development, efficacy, and controversies. J Am Acad Dermatol 69(6):S100–S109
22. Poon TSC, Barnetson RSC et al (2003) Prevention of immunosuppression by sunscreens in humans is unrelated to protection from erythema and dependent on protection from ultraviolet A in the face of constant ultraviolet B protection. J Invest Dermatol 121(1):184–190
23. Melnikova VO, Ananthaswamy HN (2005) Cellular and molecular events leading to the development of skin cancer. Mutat Res 571(1–2):91–106
24. Fourtanier A, Moyal D et al (2005) Measurement of sunscreen immune protection factors in humans: a consensus paper. J Invest Dermatol 125(3):403–409
25. Ullrich SE, Kim TH et al (1999) Sunscreen effects on UV-induced immune suppression. J Investig Dermatol Symp Proc 4(1):65–69

26. Inadequate SPF demands better UVA protection. https://www.dermatologytimes.com/view/ina
    dequate-spf-demands-better-uva-protection. Accessed 22 Feb 2024
27. Bode AM, Roh E. Are FDA-approved sunscreen components effective in preventing solar
    UV-induced skin cancer? Cells 9(7):1674
28. Schilling K, Bradford B et al (2010) Human safety review of "nano" titanium dioxide and zinc
    oxide. Photochem Photobiol Sci 9(4):495–509
29. Over-the-counter sunscreen drug products; required labelling based on effectiveness testing.
    Electronic Code of Federal Regulations Title 21, Section 201.327 (21CFR201.327). https://
    www.ecfr.gov/current/title-21/chapter-I/subchapter-C/part-201. Accessed 22 Feb 2024
30. Using TiO_2 and ZnO for balanced UV protection. https://www.personalcaremagazine.com/
    story/5243/using-TiO2-and-zno-for-balanced-uv-protection. Accessed 22 Feb 2024
31. Reinosa JJ, Leret P et al (216) Enhancement of UV absorption behavior in ZnO–TiO_2
    composites. Boletín Sociedad Española Cerámica Vidrio 55(2):55–62
32. Reinosa JJ, Docio CMÁ et al (2018) Hierarchical nano ZnO-micro TiO_2 composites: high UV
    protection yield lowering photodegradation in sunscreens. Ceram Int 44(3):2827–2834
33. Wang SQ, Tooley IR (2011) Photoprotection in the era of nanotechnology. Semin Cutan Med
    Surg 30(4):210–213
34. Ghamarpoor R, Fallah A et al (2023) Investigating the use of titanium dioxide (TiO_2)
    nanoparticles on the amount of protection against UV irradiation. Sci Rep 13(1):9793
35. Ilić K, Selmani A et al (2020) The shape of titanium dioxide nanomaterials modulates their
    protection efficacy against ultraviolet light in human skin cells. J Nanopart Res 22:71
36. Lademann J, Schanzer S et al (2005) Synergy effects between organic and inorganic UV filters
    in sunscreens. J Biomed Opt 10(1):014008
37. The synergistic benefits of using organic and inorganic UV filters in sun care. https://cosmetics
    business.com/the-synergistic-benefits-of-using-organic-and-inorganic-uv-filters-in-sun-care--
    170450. Accessed 22 Feb 2024
38. Ginzburg AL, Blackburn RS et al (2021) Zinc oxide-induced changes to sunscreen ingredient
    efficacy and toxicity under UV irradiation. Photochem Photobiol Sci 20(10):1273–1285
39. Popov AP, Lademann J et al (2005) Effect of size of TiO_2 nanoparticles embedded into stratum
    corneum on ultraviolet-A and ultraviolet-B sun-blocking properties of the skin. J Biomed Opt
    10(2005):064037
40. Lewicka ZA, Benedetto AF et al (2011) The structure, composition, and dimensions of TiO_2
    and ZnO nanomaterials in commercial sunscreens. J Nanopart Res 13:3607–3617
41. Diaz JH, Nesbitt LT (2013) Sun exposure behavior and protection: Recommendations for
    travelers. J Travel Med 20(2):108–118
42. Diffey B (2009) Sunscreens: Expectation and realization. Photodermatol Photoimmunol
    Photomed 25(5):233–236
43. Faco HAL, Guillermo MJ et al (2022) Potential systemic toxicity of UV filters in sunscreen: a
    review. Int J Res Publ Rev 3(5):3176–3191
44. Singh N, Manshian B et al (2009) NanoGenotoxicology: the DNA damaging potential of
    engineered nanomaterials. Biomaterials 30(23–24):3891–3914
45. Picatonotto T, Vione D et al (2001) Photocatalytic activity of inorganic sunscreens. J Dispers
    Sci Technol 22(4):381–386
46. Mancuso JB, Maruthi R et al (2017) Sunscreens: an update. Am J Clin Dermatol 18(5):643–650
47. Serpone N, Salinaro A et al (2002) An in vitro systematic spectroscopic examination of the
    photostabilities of a random set of commercial sunscreen lotions and their chemical UVB/UVA
    active agents. Photochem Photobiol Sci 1(12):970–981

Chapter 5
Human Safety of Sunscreens Containing ZnO and TiO$_2$ UV Filters

The very properties of TiO$_2$ and ZnO, which make them attractive for use as UV filters, have also raised alarms regarding their safety when applied to skin daily as actives in sunscreens. Due to their diminutive size, photoactivity, high surface area, and other properties, the entry of TiO$_2$ and ZnO particles, especially NPs, into the human body from sunscreen use through dermal penetration, inhalation, ingestion, and other mechanisms and their potential effects in the body are of concern [1–4]. Several review articles in the literature have focused on the safety of inorganic UV filters in sunscreens and their toxic effects on biological functions [4–10]. Studies have indicated that exposure to TiO$_2$ and ZnO particles, both in the absence and presence of UV radiation, can cause different harmful effects to human and animal cells through direct and indirect mechanisms [11]. The toxicity of TiO$_2$ and ZnO is primarily linked to the reactive oxygen species (ROS), mainly superoxide radicals (O$_2^{-\bullet}$) and hydroxyl radicals ($^\bullet$OH), and to a lesser extent, H$_2$O$_2$ and singlet oxygen (^{1}O$_2$), that are generated on the irradiation of TiO$_2$ or ZnO by UVR in sunlight [3, 11]. ROS are known to cause oxidative stress and inflammatory responses in cells and are largely responsible for the cytotoxic and genotoxic effects of inorganic UV filters on organisms [11]. Oxidative stress arises when a cell cannot neutralize all the ROS generated by inorganic UV filters, resulting in damage to the different components of cells [11]. Additionally, sufficiently small inorganic UV filter NPs can potentially pass through the cell membrane and cause direct effects inside cells by interacting with the contents of cell cytoplasm or the nucleus [11].

Dunford et al. [12] found that sunlight-irradiated TiO$_2$ catalysed the damage of DNA in cultured human (MRC-5 fibroblasts) cells in both in vitro and in vivo studies. In a study by Park et al. [13], the exposure of cultured human bronchial epithelial (BEAS-2B) cells to TiO$_2$ NPs resulted in increased ROS levels, decreased reduced glutathione (GSH) levels, activation of different oxidative stress-related genes and a hypoxia inducible gene, and caused cell death. The uptake of TiO$_2$ NPs into the cytoplasm of BEAS-2B cells was also observed in this study, with particles aggregating in the perinuclear region, where the particles could directly interact with the

N. H. Kera et al., *Inorganic Ultraviolet Filters in Sunscreen Products*, SpringerBriefs in Materials, https://doi.org/10.1007/978-3-031-64114-5_5

molecules in the cells and cause deleterious biological responses. Even though the focus here was on the effect of TiO$_2$ NPs on human lung epithelial BEAS-2B cells, in the absence of UVR, this study is relevant here due to the common use of aerosol sunscreens and as the effects observed could also occur in skin and other cells in the body exposed to TiO$_2$ NPs due to sunscreen use. In in vivo studies, TiO$_2$ NPs were found to cause minor pro-inflammatory effects on human endothelial cells, cytotoxic effects in rat alveolar macrophages, genotoxic effects in Syrian Hamster Embryo Fibroblasts and both genotoxic and cytotoxic effects in cultured human lymphoblastoid cells [14–17]. Oxidative damage of DNA was observed in *Saccharomyces cerevisiae* yeast in the presence of TiO$_2$ under UV irradiation [18, 19]. Liu et al. [20] found that TiO$_2$ NPs (anatase) caused greater oxidative stress and damage to the liver of mice than bulk-TiO$_2$.

Jeng et al. [21] found that ZnO exhibited cytotoxic effects on Neuro-2A cells. ZnO NPs at doses of 50–100 μg/mL were found to decrease the mitochondrial function in Neuro-2A cells significantly, and at doses above 100 μg/mL caused cell damage. In comparison, TiO$_2$ at doses below 200 μg/mL did not cause any observable effects on the cells.

Dufour et al. [22] investigated the clastogenicity, in terms of percentage of cells with chromosome aberrations, of ZnO NPs of 100 nm size, to Chinese hamster ovary cells under different UV-irradiation conditions: in the dark (with no irradiation), with UV-irradiation of cells before treatment with ZnO NPs, and with simultaneous irradiation of cells and ZnO NPs. Chromosomal aberrations caused by exposure to ZnO NPs occurred in all cases. Still, they were higher in the pre-irradiated and simultaneously irradiated cells than those kept in the dark and higher in the simultaneously-irradiated cells than the pre-irradiated cells. The lowest dose of ZnO NPs that caused a significant increase in damage to DNA was 105 mg/mL for cells kept in the dark but a much lower concentration of 54 mg/mL for pre-irradiated and simultaneously irradiated cells.

For ZnO and TiO$_2$ particles to cause the toxic and other effects in the body, observed in different in vivo and in vitro studies, some of which are detailed above, the particles would have to pass the stratum corneum, move through the dermis and enter the bloodstream [1, 2, 6, 8]. However, the consensus of findings from both in vivo and in vitro dermal penetration studies in the literature is that TiO$_2$ and ZnO NPs do not typically penetrate the stratum corneum layer of normal and intact skin [1, 2, 6, 8–10]. However, in vivo and ex vivo studies have indicated that low levels of zinc ions, arising from the dissolution of ZnO NPs or zinc salts within different formulation mediums, can penetrate the deeper layers of the skin when applied as ZnO NPs [24–28].

Sadrieh et al. [29] investigated the dermal penetration of uncoated TiO$_2$ particles (nano- and submicron-sized) and coated TiO$_2$ NPs (dimethicone/methicone copolymer-coated) in sunscreen formulations (5 wt%) in minipigs. TiO$_2$ was detected as titanium in the epidermis and neck and abdominal dermis of minipigs treated with coated and uncoated nanosized TiO$_2$. However, there was no significant dermal penetration by micro- and nanosized TiO$_2$ in formulations topically applied to minipigs for 1 month. Previous in vitro tests carried out on porcine skin, due to its likeness to human skin, showed similar findings. Microfine and nanosized TiO$_2$ and ZnO particles and their respective ions in sunscreen formulations did not penetrate the stratum corneum of pigs in in vitro studies carried out by Gamer et al. [30]. Wu et al. [31] found in in vitro studies that TiO$_2$ NPs of sizes 4 and 60 nm did not penetrate the epidermis of pig ears. However, in in vivo studies, the TiO$_2$ NPs were found to penetrate the skin of hairless mice, move into tissues, and cause lesions in different organs in the body. This was attributed to the different properties of hairless mice skin when compared to that of porcine skin, such as lower thickness and barrier properties, which may enhance NP penetration, and led to questions raised regarding the suitability of hairless mice as a model for human skin and relevance of the results obtained.

Mohammed et al. [32] concluded from in vivo studies conducted on healthy human skin that uncoated and coated ZnO NPs did not penetrate beyond the stratum corneum. ZnO-NPs were found to accumulate on the surface and in furrows of the skin but did not penetrate the epidermis or cause toxic effects on cells or apoptosis. Dispersions of commercial uncoated ZnO NPs (Z-Cote) and triethoxycaprylylsilane-coated ZnO NPs (Z-cote HP1) in caprylic/capric triglyceride were applied to the skin of human subjects, with no skin damage or disease, according to an hourly and daily schedule. Multiphoton tomography and fluorescence lifetime imaging microscopy were used to observe ZnO NPs on the skin and assess cell health. Labile zinc species from ZnO NPs on excised human skin were slightly higher than those of the control group.

Khabir et al. [26] investigated the distribution and levels of ZnO NPs and zinc ions in excised human skin samples topically treated with sunscreen formulations containing ^{67}ZnO NPs, containing the zinc-67 isotope (^{67}Zn). The study found that ^{67}ZnO NPs remained on the skin surface and did not penetrate further than the superficial layers of the stratum corneum, even after 5 days of repeated application. The higher levels of intracellular and labile zinc in the stratum corneum and viable epidermis of treated skin than that of the control indicated the penetration of zinc ions, not ZnO NPs, across the stratum corneum and into the viable epidermis. However, the concentrations of absorbed ^{67}Zn in the viable epidermis were found to be lower than the endogenous zinc levels there, and the total zinc concentrations in the viable epidermis were lower than the levels that cause toxic effects to human keratinocyte cells (HaCaT).

A study by Pelclova et al. [33] conducted on human subjects, in which sunscreens containing TiO$_2$ NPs were found to prevent erythema caused by exposure to UVR but did not decrease the levels of biomarkers that indicate oxidative stress and inflammation caused by UVR exposure. The researchers suggested that the detection of TiO$_2$ NPs in urine and plasma, but not in exhaled breath condensate, suggested skin penetration by TiO$_2$ NPs. However, analysis of local skin samples was not conducted to corroborate this finding. Furthermore, this study used only a small sample size of six test subjects, necessitating further investigation.

Limsakul et al. [34] compared the skin penetration propensity of one-dimensional (1D) TiO$_2$ nanowires to that of commercial spherical TiO$_2$ NPs via in vitro studies carried out on porcine skin. 1D TiO$_2$ nanowires were obtained by calcining the

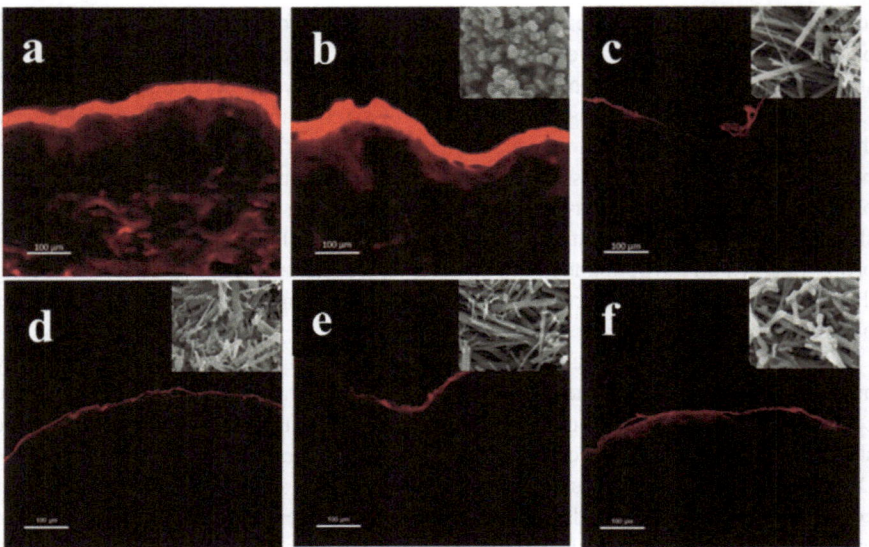

Fig. 5.1 CLSM images obtained of porcine skin sections treated with **a** Nile red dye, **b** Nile red-tagged spherical TiO₂ NPs, and Nile red-tagged 1D TiO₂ nanowires, **c** Pre-calcined, and calcined at **d** 400 °C, **e** 700 °C and **f** 900 °C, with SEM images of samples shown in top right corners of images. Reproduced with permission from [34], with slight modifications. Copyright 2023, Elsevier Science Ltd.

product of a modified hydrothermal synthesis method at three different temperatures. A fluorescent dye, Nile red, was used to investigate skin penetration by different TiO₂ particles. Imaging was carried out on a confocal laser scanning microscope (CLSM) to compare the skin penetration of Nile red only (the control) to that of Nile-red tagged TiO₂ particles. The images obtained (Fig. 5.1) showed that Nile red-tagged spherical TiO₂ NPs (20–30 nm in size) penetrated deeper layers of the skin, similar to that of the control (only Nile red dye applied).

In contrast, the Nile red-tagged 1D TiO₂ nanowire samples were only localized on the upper epidermis layer. The synthesized different 1D TiO₂ nanowires had widths ranging between 122 and 199 nm and lengths ranging between 238 and 670 nm. The authors suggested that particle shape was a major factor influencing the skin penetration of particles. However, the effect of particle size on particle penetration into the skin was not delved into, and the significantly smaller size of the spherical TiO₂ NPs compared to that of the 1D TiO₂ nanowires makes true comparison difficult.

The potential of NPs to penetrate compromised skin has also raised concerns [23]. However, the results from several studies have largely indicated insignificant dermal penetration of TiO₂ and ZnO NPs through sunburnt and/or damaged porcine and human skin [23]. Furthermore, certain skin conditions, such as psoriasis vulgaris, could even decrease dermal penetration of sunscreen products applied to the skin [23].

Pinheiro et al. [35] found that TiO$_2$ NPs were not significantly absorbed and did not penetrate normal and psoriatic human skin. While TiO$_2$ NPs remained on the outer layers of the stratum corneum of normal skin, the frailty of the stratum corneum of psoriatic skin and the corneocytes arrangement permitted slight TiO$_2$ NP penetration. Still, these NPs did not reach the layers below. Sufficient distribution of cream on the dermal surface of psoriatic skin was hampered due to the desquamation on the stratum corneum.

Filipe et al. [36] investigated the dermal penetration of different sunscreen formulations, a hydrophobic test formulation containing coated TiO$_2$ NPs and two commercial formulations containing only TiO$_2$ NPs or a combination of TiO$_2$ and ZnO, in in vivo studies on human subjects. Dermal penetration of TiO$_2$ and/or ZnO was investigated after 2 h exposure periods on normal healthy skin, psoriatic skin and artificially damaged skin, under occlusion. Nuclear microscopy analysis showed that TiO$_2$ and/or ZnO NPs were present on the skin surface and the upper layers of the stratum corneum and deposited preferentially in the pilosebaceous follicles openings but were not detected beyond the stratum corneum in any of the cases. Similar findings were reported in studies by Gulson et al., Monteiro-Riviere et al., and Gopee et al. [28, 37, 38]. In a study by Næss et al. [39] a sunscreen formulation containing coated TiO$_2$ NPs was applied six times per day for 7 days to the skin on the backs of two human subjects; TEM–EDX analysis indicated the presence of titanium-containing particles below the stratum corneum, in the strata granulosum and spinosum layers, in both intact skin and UVB-damaged skin, prior and post-exposure to UVB radiation. Holmes et al. [27] investigated the penetration of zinc into intact and impaired human skin under real-use conditions by applying formulations of ZnO NPs in artificial sweat or caprylic capric triglyceride, a typical sunscreen medium, onto ex vivo skin. The study found that the increase in zinc concentrations in the skin strata depended on different factors, including the skin barrier (stratum corneum) integrity, application duration and, to a greater extent, the UV filter medium. Zinc levels in intact skin significantly increased over time after the application of ZnO NPs formulations in artificial human sweat, suggesting the penetration of zinc ions arising from the dissolution of ZnO NP in human sweat of acidic pH (~4.8). Zn levels in intact skin also increased after application of ZnO NP formulations in caprylic capric triglyceride medium, and this was attributed to the hydration of the stratum corneum facilitating ZnO dissolution. The increase in zinc levels in the viable epidermis was higher in impaired skin than in intact skin. Zinc concentrations in impaired skins increased by 60–65% and over 100%, compared to the control, for both the sweat and caprylic capric triglyceride mediums, after 24 h and 48 h, respectively. Cytotoxic and genotoxic effects arising from different types of physicochemical interactions (particle–skin, particle–particle, and skin–particle–light) involving ZnO or TiO$_2$ particles deposited in the stratum corneum of the skin due to sunscreen use are also of concern and require investigation [3].

The main risk associated with exposure to inorganic NPs/particles is through the inhalation route, as the lungs cannot clear the particles that result there [7]. The use of powdered or aerosol sunscreen products can expose the lungs to inorganic UV filters [40]. Sayes et al. [41] assessed the pulmonary toxicity of fine-sized (<1000 nm) and nanosized ZnO (50–70 nm) and other inorganic particles, including carbonyl iron (CI) (800–3000 nm), crystalline silica (CS) (1600 nm), and precipitated amorphous silica (AS) (1000–3000 nm), in both in vitro and in vivo studies. In the in vivo studies, rats were exposed to a 1–5 mg/kg dose of particles through intratracheal installation. The lungs were lavaged for measurement of markers for cytotoxicity endpoints, in terms of bronchoalveolar lavage fluid lactate dehydrogenase (LDH) values, and inflammation, in terms of neutrophil recruitment, at 24 h, 1 week, 1 month, and 3 months after exposure. In the in vitro studies, the different particles investigated, at varying doses, were incubated with three different cell cultures: rat L2 lung epithelial cells, primary alveolar macrophages (AMs) (collected through bronchoalveolar lavage from unexposed rats), and AM—L2 lung epithelial cell cocultures. The cell culture fluids were assessed for inflammatory cytokines (macrophage inflammatory 2 protein [MIP-2], tumour necrosis factor alpha [TNF-α], and interleukin-6 [IL-6]) and cytotoxicity endpoints (LDH, 1-(4,5-dimethylthia-zol-2-yl)-3,5-diphenylformazan [MTT]) at different times. The results of in vivo studies indicated that inorganic particles showed low toxicity in the lungs of rats and that the observed inflammatory responses were reversible. The results of in vitro studies varied for the different particles investigated. L2 cells showed the highest susceptibility to the cytotoxic effects of inorganic particles, with the fine-sized and nanosized ZnO particles showing higher cytotoxicity to these cells than crystalline silica or amorphous silica for 4 or 24 h exposure times. Epithelial macrophage cocultures were less sensitive, and macrophages showed high resistance to the effects of inorganic particles. However, the authors concluded that there was little correlation between the results obtained from the in vivo and in vitro studies under the experimental conditions in this study. Furthermore, effects were only observed for high particle doses, typically used to check worst-case scenario exposure.

In another study, Sayes et al. [42] found that the cytotoxicity of TiO₂ NPs towards human dermal fibroblasts (HDF) and human lung epithelial (A549) cells correlated well with their photoactivity, largely phase-dependent, towards the photodegradation of aqueous solution of Congo Red. The highly photoreactive TiO₂ phase, anatase, displayed higher cytotoxicity, by two orders of magnitude, than rutile NPs, which exhibited significantly lower inherent photoactivity. Anatase TiO₂ NPs, at sufficiently high concentrations, affected the normal cellular activity of HDF and A549 cells as indicated by increased LDH release, decreased metabolic mitochondrial activity, and increased production of interleukin, an inflammatory mediator, in a dose- and time-response manner and the responses increased significantly with UV illumination. However, cytotoxicity and inflammation responses were only observed for relatively high doses of anatase TiO₂ NPs, 1500 μg/mL for increased LDH release and decreased metabolic mitochondrial activity, and 300 μg/mL for increased production of interleukin. This study suggested that TiO₂ phases that generated high levels of ROS on UV–irradiation in photocatalytic studies were more predisposed towards the generation of ROS in biological systems, mainly responsible for cytotoxic effects in cell cultures. It must also be noted that TiO₂ has been classified as a group 2B carcinogen by the International Agency for Research on Cancer (IARC), deeming it a possible carcinogen to humans [3]. This conclusion was based on research findings that indicated that exposure to high concentrations of TiO$_2$, of both pigment grade and ultrafine particle sizes, through the inhalation route caused respiratory tract cancer in rats. The research was considered relevant to humans as the biological events observed in these studies on rat models also occurred in human subjects who encountered TiO$_2$ dust through occupational exposure by inhalation.

A well-known disadvantage associated with the use of organic or chemical UV filters in sunscreens and other products is the adverse effects caused by simultaneous exposure to two or more organic filters, such as allergic and photoallergic contact dermatitis and the development of hypersensitivity in certain individuals to one or more UV filters, often from different chemical classes [43]. Even though this is not a challenge typically associated with the use of inorganic UV filters, the development of possibly adverse effects on the combination of two or more types of inorganic filters or inorganic UV filters with organic UV filters, a strategy commonly employed for achieving products with broad-spectrum UV protection, has received attention.

Hackenberg et al. [44] investigated the cytotoxic and genotoxic effects of TiO$_2$ NPs and ZnO NPs, individually, and when combined. Cultured human nasal mucosa cells were exposed to ZnO NPs, TiO$_2$ NPs, and combinations of the two at concentrations ranging from 0.1 to 20 μg/ml for 24 h in a battery of tests. The comet assay was used to measure genotoxic effects, such as damage to DNA and the capacity for DNA repair, while the MTT assay was used to assess cytotoxic effects. The results obtained showed that cytotoxic and genotoxic effects were observed in cells exposed to ZnO NPs, which supported the results of an earlier in vitro study by the same group. Still, these effects were not observed for TiO$_2$ at the concentrations tested in this study [44, 45]. In the presence of TiO$_2$ NPs, the genotoxic effects induced by ZnO NPs in cells were found to decrease significantly, and the capacity of cells to repair DNA damage due to contact with ZnO NPs was enhanced. The antagonization of the genotoxic effects of ZnO by TiO$_2$ NPs was attributed to decreased zinc ion concentrations due to their adsorption by TiO$_2$ NPs. This study suggested that the combination of TiO$_2$ and ZnO NPs in sunscreens could increase their safety for use and may offer additional advantages, such as the use of lower concentrations of the NPs. However, this study did not investigate the UV-attenuation properties and safety of the individual NPs and NP-combinations when incorporated into sunscreen formulations.

Concerns have also been raised over Vitamin D deficiency arising due to sunscreen use, as exposure to sunlight is essential for the synthesis of Vitamin D in the skin and accounts for over 90% of the Vitamin D in the body [46–48]. The absorption of UVB radiation in sunlight by 7-dehydrocholesterol in the skin brings about its conversion to previtamin D3, which undergoes isomerization to form the biologically active vitamin D3. Metabolization of Vitamin D3 in the liver results in its circulating form, 25-hydroxyvitamin D, which in turn is metabolized in the kidneys to its biologically active form, 1,25-dihydroxyvitamin D, which plays an important role in maintaining bone health by regulating the metabolism of phosphate and calcium [46–48]. Vitamin D deficiency has been linked to an increased risk of bone fractures, diabetes, cardiovascular disease, cancer, depression, cognitive decline, and death.

Studies have reported decreased Vitamin D levels associated with sunscreens when applied rigorously and at the recommended dose used for SPF testing, i.e. 2 mg/cm^2. However, further studies concluded that Vitamin D deficiency was not likely to arise from sunscreen use as sunscreens are typically applied at a much lower dose (1.5 mg/cm^2) than that recommended and may be linked to other sun protection practices such as wearing protective clothing and staying in the shade [46–48]. Vitamin D deficiency may also be caused by factors other than insufficient sunlight exposure, such as genetics and diet. Furthermore, adequate Vitamin D levels in the body can be achieved through sensible sun exposure, such as exposure to sunlight for a duration that is roughly half of that which would cause mild sunburn or, in the case of people at risk for Vitamin D insufficiency, by taking a Vitamin D supplement.

Studies in the literature have indicated that metal oxide UV filter particles can interact with human and animal cells and can cause different adverse effects under conditions with or without UV irradiation, primarily due to generated ROS. However, most dermal penetration studies have shown that TiO$_2$ and ZnO NPs do not penetrate both normal and/or intact skin, as well as sunburnt and/or damaged human and animal skin. Therefore, the likelihood of metal oxide particles from topically applied sunscreens interacting with and causing adverse effects in cells is low. As the main risk of exposure to TiO$_2$ and ZnO is to the lungs via inhalation, aerosol and spray-based sunscreen formulations should be avoided. The limitations of the studies carried out to investigate the toxicity of ZnO, TiO$_2$, and other particles used as UV filters must also be noted [35]. The in vitro and in vivo studies employed are often not standardized, making comparisons difficult between different studies. The lack of correlation of the results obtained from in vitro studies with that of in vivo studies has led to questions regarding the relevance of in vitro studies and the extent of the influence of the selection of different experimental variables such as cell types, particle dose, exposure duration, and time course, among others, on the results obtained. For example, the selection of an appropriate particle dose for toxicology studies is critical for obtaining meaningful results. However, the high particle doses selected for many in vitro and in vivo studies, typically higher than the levels of inorganic particles expected to occur in cells, can induce effects that may not be physiologically relevant. As such, studies have cited the need for better cell cultures, research for the determination of relevant biological, chemical, and physiological endpoints, and the development and validation of cost-effective tests for in vitro studies.

References

1. Egambaram OP, Pillai SK et al (2020) Materials science challenges in skin UV protection: a review. Photochem Photobiol 96(4):779–797
2. Manaia EB, Kaminski RCK et al (2013) Inorganic UV filters. Braz J Pharm Sci 49(2):201–209
3. Smijs TG, Pavel S et al (2011) Titanium dioxide and zinc oxide NPs in sunscreens: focus on their safety and effectiveness. Nanotechnol Sci Appl 4:95–112

4. Paiva JP, Diniz RR et al (2020) Insights and controversies on sunscreen safety. Crit Rev Toxicol 50(8):707–723
5. Adler BL, DeLeo VA (2020) Sunscreen safety: a review of recent studies on humans and the environment. Curr Dermatol Rep 9:1–9
6. Gilbert E, Pirot F et al (2013) Commonly used UV filter toxicity on biological functions: review of last decade studies. Int J Cosmet Sci 35(3):208–219
7. Ballestín SS, Bartolomé MJL (2023) Toxicity of different chemical components in sun cream filters and their impact on human health: a review. Appl Sci 13(2):712
8. Schneider SL, Lim HW (2019) A review of inorganic UV filters zinc oxide and titanium dioxide. Photodermatol Photoimmunol Photomed 35(6):442–446
9. Schilling K, Bradford B et al (2010) Human safety review of "nano" titanium dioxide and zinc oxide. Photochem Photobiol Sci 9(4):495–509
10. Fajzulin I, Zhu X et al (2015) Nanoparticulate inorganic UV absorbers: a review. J Coat Technol Res 12(4):617–632
11. Singh N, Manshian B et al (2009) NanoGenotoxicology: the DNA damaging potential of engineered nanomaterials. Biomaterials 30(23–24):3891–3914
12. Dunford R, Salinaro A et al (1997) Chemical oxidation and DNA damage catalysed by inorganic sunscreen ingredients. FEBS Lett 418(1–2):87–90
13. Park EJ, Yi J et al (2008) Oxidative stress and apoptosis induced by titanium dioxide nanoparticles in cultured BEAS-2B cells. Toxicol Lett 180(3):222–229
14. Afaq F, Abidi P et al (1998) Cytotoxicity, pro-oxidant effects and antioxidant depletion in rat lung alveolar macrophages exposed to ultrafine titanium dioxide. J Appl Toxicol 18(5):307–312
15. Rahman Q, Lohani M et al (2002) Evidence that ultrafine titanium dioxide induces micronuclei and apoptosis in Syrian hamster embryo fibroblasts. Environ Health Perspect 110(8):797–800
16. Wang JJ, Sanderson BJS et al (2007) Cyto- and genotoxicity of ultrafine TiO_2 particles in cultured human lymphoblastoid cells. Mutat Res Genet Toxicol Environ 628(2):99–106
17. Peters K, Unger RE et al (2004) Effects of nano-scaled particles on endothelial cell function in vitro: studies on viability, proliferation and inflammation. J Mater Sci Mater Med 15(4):321–325
18. Pinto AV, Deodato EL et al (2010) Enzymatic recognition of DNA damage induced by UVB-photosensitized titanium dioxide and biological consequences in Saccharomyces cerevisiae: evidence for oxidatively DNA damage generation. Mutat Res 688(1–2):3–11
19. Paiva JP, Santos BAMC et al (2014) Titanium dioxide–montmorillonite nanocomposite as photoprotective agent against ultraviolet B radiation-induced mutagenesis in Saccharomyces cerevisiae: a potential candidate for safer sunscreens. J Pharm Sci 103(8):2539–2545
20. Liu H, Ma L et al (2010) Toxicity of nano-anatase TiO_2 to mice: liver injury, oxidative stress. Toxicol Environ Chem 92(1):175–186
21. Jeng HA, Swanson J (2006) Toxicity of metal oxide nanoparticles in mammalian cells. J Environ Sci Health A Tox Hazard Subst Environ Eng 41(12):2699–2711
22. Dufour EK, Kumaravel T et al (2006) Clastogenicity, photo-clastogenicity or pseudo-photo-clastogenicity: genotoxic effects of zinc oxide in the dark, in pre-irradiated or simultaneously irradiated Chinese hamster ovary cells. Mutat Res Genet Toxicol Environ Mutagen 607(2):215–224
23. Nohynek GJ, Dufour EK (2012) Nano-sized cosmetic formulations or solid nanoparticles in sunscreens: a risk to human health? Arch Toxicol 86(7):1063–1075
24. Holmes AM, Song Z et al (2016) Relative penetration of zinc oxide and zinc ions into human skin after application of different zinc oxide formulations. ACS Nano 10(2):1810–1819
25. Pirot F, Millet J et al (1996) In vitro study of percutaneous absorption, cutaneous bioavail-ability and bioequivalence of zinc and copper from five topical formulations. Skin Pharmacol 9(4):259–269
26. Khabir Z, Holmes AM et al (2021) Human epidermal zinc concentrations after topical application of ZnO nanoparticles in sunscreens. Int J Mol Sci 22:12372
27. Holmes AM, Kempson I et al (2020) Penetration of zinc into human skin after topical application of nano zinc oxide used in commercial sunscreen formulations. ACS Appl Bio Mater 3(6):3640–3647

28. Gulson B, Mccall M et al (2010) Small amounts of zinc from zinc oxide particles in sunscreens applied outdoors are absorbed through human skin. Toxicol Sci 118(1):140–149
29. Sadrieh N, Wokovich AM et al (2010) Lack of significant dermal penetration of titanium dioxide from sunscreen formulations containing nano– and submicron–size TiO$_2$ particles. Toxicol Sci 115(1):156–166
30. Gamer AO, Leibold E et al (2006) The in vitro absorption of microfine zinc oxide and titanium dioxide through porcine skin. Toxicol in Vitro 20(3):301–307
31. Wu J, Liu W et al (2009) Toxicity and penetration of TiO$_2$ nanoparticles in hairless mice and porcine skin after subchronic dermal exposure. Toxicol Lett 191(1):1–8
32. Mohammed YH, Holmes A et al (2019) Support for the safe use of zinc oxide nanoparticle sunscreens: lack of skin penetration or cellular toxicity after repeated application in volunteers. J Invest Dermatol 139(2):308–315
33. Pelclova D, Navratil T et al (2019) NanoTiO$_2$ sunscreen does not prevent systemic oxidative stress caused by UV radiation and a minor amount of NanoTiO$_2$ is absorbed in humans. Nanomaterials (Basel) 9(6):888
34. Limsakul S, Mahatnirunkul T et al (2023) Novel physical sunscreen from one-dimensional TiO$_2$ nanowire: Synthesis, characterization and the effects of morphologies and particle size for use as a physical sunscreen. Nano-Struct Nano-Objects 35:101027
35. Pinheiro T, Pallon J et al (2007) The influence of corneocyte structure on the interpretation of permeation profiles of nanoparticles across skin. Nucl Instrum Methods Phys Res B 260(1):119–123
36. Filipe P, Silva JN et al (2009) Stratum corneum is an effective barrier to TiO$_2$ and ZnO nanoparticle percutaneous absorption. Skin Pharmacol Physiol 22(5):266–275
37. Monteiro-Riviere NA, Wiench K et al (2011) Safety evaluation of sunscreen formulations containing titanium dioxide and zinc oxide nanoparticles in UVB sunburned skin: an in vitro and in vivo study. Toxicol Sci 123(1):264–280
38. Gopee NV, Roberts DW et al (2007) Migration of intradermally injected quantum dots to sentinel organs in mice. Toxicol Sci 98(1):249–257
39. Næss EM, Hofgaard et al (2016) Titanium dioxide nanoparticles in sunscreen penetrate the skin into viable layers of the epidermis: a clinical approach. Photodermatol Photoimmunol Photomed 32(1):48–51
40. Dréno B, Alexis A et al (2019) Safety of titanium dioxide nanoparticles in cosmetics. J Eur Acad Dermatol Venereol 33(S7):34–46
41. Sayes CM, Reed KL et al (2007) Assessing toxicity of fine and nanoparticles: comparing in vitro measurements to in vivo pulmonary toxicity profiles. Toxicol Sci 97(1):163–180
42. Sayes CM, Wahi R et al (2006) Correlating nanoscale titania structure with toxicity: a cytotoxicity and inflammatory response study with human dermal fibroblasts and human lung epithelial cells. Toxicol Sci 92(1):174–185
43. Uter W, Gonçalo et al (2014) Coupled exposure to ingredients of cosmetic products: III. Ultraviolet filters. Contact Dermatitis 71(3):162–169
44. Hackenberg S, Scherzed A et al (2017) Genotoxic effects of zinc oxide nanoparticles in nasal mucosa cells are antagonized by titanium dioxide nanoparticles. Mutat Res Genet Toxicol Environ Mutagen 816–817:32–37
45. Hackenberg S, Scherzed A et al (2011) Cytotoxic, genotoxic and pro-inflammatory effects of zinc oxide nanoparticles in human nasal mucosa cells in vitro. Toxicol in Vitro 25(3):657–663
46. Serpone N (2021) Sunscreens and their usefulness: have we made any progress in the last two decades? Photochem Photobiol Sci 20(2):189–244
47. Wacker M, Holick MF (2013) Sunlight and Vitamin D: a global perspective for health. Dermatoendocrinol. 5(1):51–108
48. Lim HW, Arellano-Mendoza MI et al (2017) Current challenges in photoprotection. J Am Acad Dermatol 76(3S1):S91–S99

Chapter 6
Environmental Safety and Impact of Sunscreens Containing Inorganic UV Filters

An inevitable negative consequence of the use of sunscreens for photoprotection is the pollution of the environment by UV filters. The occurrence of inorganic UV filters in the environment is highly concerning due to their properties, such as small sizes, chemistry, reactivity, photoreactivity, and propensity to bioaccumulate. The sources, fate, and potential effects of inorganic UV filters on organisms in the environment have, therefore, been extensively investigated and are the subject of several recent review articles in the literature [1–5]. The different aspects related to inorganic UV filters released in the environment, such as potential sources, interactions, and toxicity, are summarized in Fig. 6.1a [4].

Inorganic UV filters typically end up in the ocean and other water bodies directly due to wash-off during recreational activities such as swimming and occur in greywater from baths, showers, hand basins, and washing machines due to their use in topically applied products such as sunscreens, personal care products and cosmetics [2–5]. Inorganic UV filters can also result in the environment indirectly due to the release of effluents from wastewater treatment plants. Inorganic UV filters have been detected in surface waters, groundwater, wastewater treatment plant effluents, sediments, aquatic plants, and animals. The quantification of ZnO and TiO_2 levels arising solely from UV filters in the environment is difficult due to their natural occurrence there and use in a wide range of products and applications and is further complicated due to complex interactions arising due to particle properties and environmental factors [3, 5–7]. However, the concentrations of ZnO and TiO_2 in aquatic bodies were measured and/or estimated to be very low (as low as 10 μg/L) in different studies [3, 6, 7]. On release into aqueous environments, ZnO and TiO_2 inorganic UV filters undergo different physical changes such as dissolution, dispersion, aggregation, sedimentation, adsorption, and bioaccumulation in biota and cause effects to different extents depending on their form and properties [3, 5–7]. Studies have shown that ZnO UV filters quickly undergo dissolution in water, forming hydrated Zn^{2+} cations. At the same time, the more stable TiO_2, of low aqueous solubility, tends to form aggregates that suspend in water or undergo sedimentation [2–5]. Even though there

N. H. Kera et al., *Inorganic Ultraviolet Filters in Sunscreen Products*, SpringerBriefs in Materials, https://doi.org/10.1007/978-3-031-64114-5_6

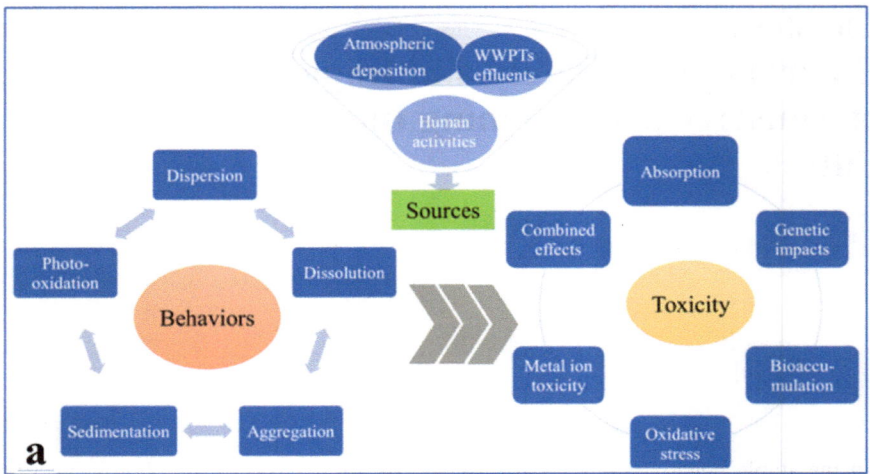

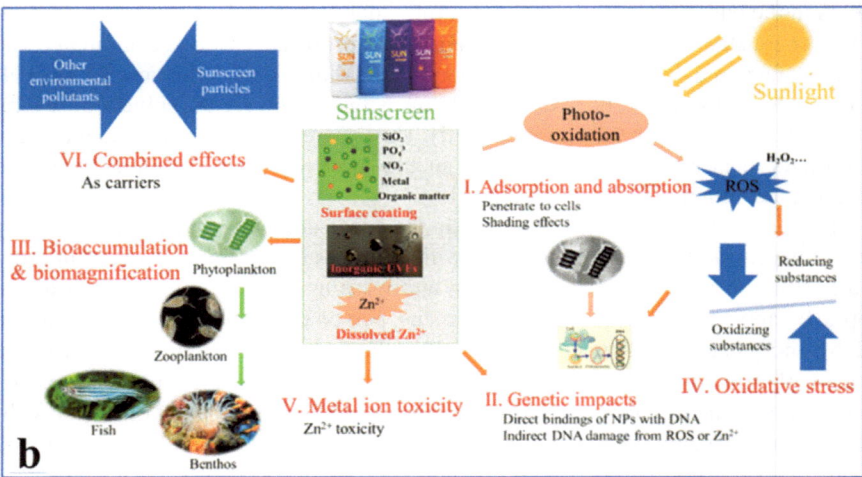

Fig. 6.1 Schemes showing potential **a** sources, interactions, and toxicity, and **b** fate, effects and toxicity mechanisms of inorganic UV filters in the environment. Reproduced with permission from [4], with slight modifications. Copyright 2022, MDPI

are limited studies on the ecological impacts of inorganic UV filters, reports in the literature have indicated the potential effects of TiO_2 and ZnO on different freshwater and marine biota [1–4]. The toxicity of ZnO and TiO_2 UV filters towards aquatic organisms is due to their inherent photoreactivity. It is mainly linked to the oxidative stress induced in cells by ROS generated on the irradiation of these UV filters [8, 9]. Generated ROS can cause cell damage in different ways, such as by promoting protein oxidation that leads to protein carbonylation and impairing membrane lipids, resulting in lipid peroxidation [9]. The toxicity of metal oxide-based UV filters has also been linked to the accumulation of free metal cations arising from their dissolution in aquatic environments [9]. Figure 6.1b summarizes the potential fate, effects and toxicity mechanisms of inorganic UV filters that result in the environment [4].

There has been increasing concern over the adverse impacts of inorganic UV filters from sunscreen products on coral reefs due to their important ecological role in marine environments and their high sensitivity to even slight changes in environmental conditions [1, 2, 4, 10–13]. A literature review article by Neto et al. [10] focusing on the toxicity of ZnO UV filters to corals and zooxanthellae concluded that though studies have shown that ZnO NPs and zinc ions could cause toxic effects to certain coral and zooxanthellae species, the experimental parameters used in different studies varied significantly making scientific consensus difficult. The authors emphasized the need for validated and standardized methodologies for toxicology studies [10].

Corinaldesi et al. [14] investigated the effects of different ZnO NPs and commercially available TiO_2 NPs on a stony coral species, *Acropora spp.*, in laboratory studies. The uncoated ZnO NPs caused the rapid and severe bleaching (>60%) of *Acropora spp.* stony coral, as shown in Fig. 6.2, through altering the symbiotic interactions between zooxanthellae microalgae and the stony coral species and also triggered microbial enrichment of the seawater around the corals. This was a result of ZnO directly decreasing the populations of zooxanthellae microalgae. In contrast, the corals exposed to the two types of modified TiO_2 studied, Optisol™ (manganese-modified TiO_2) and Eusolex® T2000 (alumina and dimethicone-coated TiO_2), did not show any bleaching and microbial enrichment of the surrounding seawater was also not observed, as shown in Fig. 6.2, suggesting that modified TiO_2 was more environmentally compatible than uncoated ZnO. A limitation of this study was that unmodified TiO_2 and modified ZnO particles were not investigated, making a thorough comparison difficult. Another limitation was that only short-term exposure of corals to inorganic UV filters was investigated, and the effects of chronic and long-term exposure were not considered, which may be more relevant as inorganic UV filters tend to persist and accumulate in the marine environment.

Blaise et al. [15] assessed the ecotoxicity of different metal oxide NPs by dispersing them in an aqueous solution and subjecting the filtrates obtained to a series of bioassays covering different taxonomic groups. The amount of TiO_2 NPs retained in the filtrate, after filtration through a 0.22 μm membrane, was 54%, suggesting that NPs may be removed from water in the environment by partitioning on suspended solids and later form part of the sediment. TiO_2 NPs were found to cause toxic effects to fish cells (trout primary hepatocyte) at 1–10 mg/L levels and were harmful to the invertebrate *H. attenuate* at 10–100 mg/L levels but did not exhibit toxicity to the invertebrate *T. platyurus* and the bacteria and phototrophic systems examined.

Microalgae, or phytoplankton, are primary producers that serve a vital role in aquatic environments [2–4]. Microalgae are abundant in fresh and marine water bodies and are relatively easy to culture and grow. Therefore, they are often used as model aquatic organisms in toxicity studies. Miller et al. [16] investigated the effect of ZnO and TiO_2 NPs on the growth of four marine plankton species under low levels of

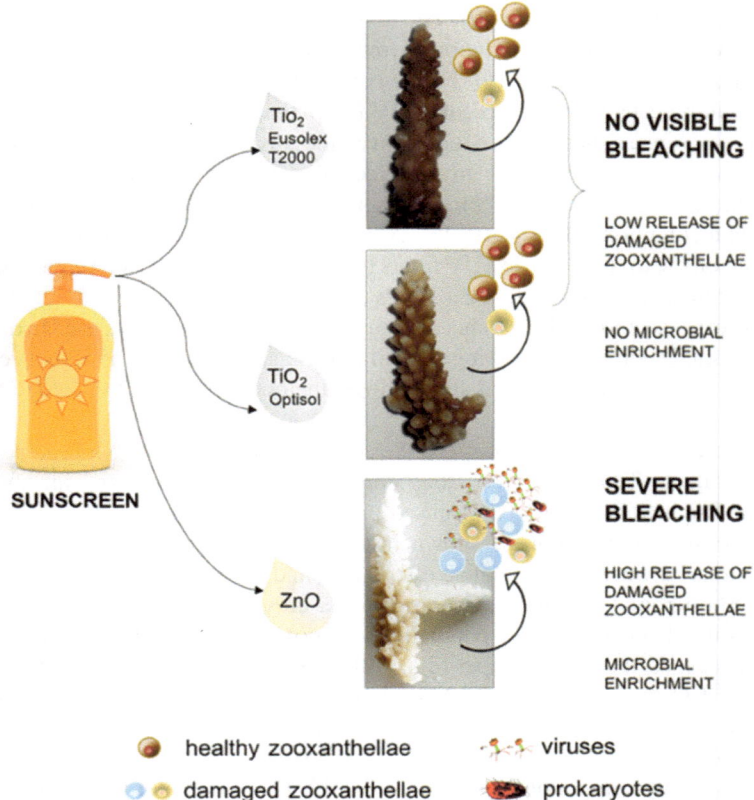

Fig. 6.2 Effects of ZnO NPs and commercially obtained TiO$_2$ NPs on the stony coral species, *Acropora spp.*, in laboratory studies. Reproduced with permission from [14], with slight modifications. Copyright 2018, Elsevier Science Ltd.

UV light from cool white fluorescent lights. The growth rates of the different marine plankton species were significantly decreased by exposure to ZnO NPs but were not affected by TiO$_2$ NPs, for NP concentrations of between 10 and 1000 µg/L. It was suggested that toxic effects were caused by free zinc ions assimilated by marine phytoplankton released from the dissolution of ZnO NPs. However, the effects of ZnO NPs could not be distinguished from those of the released zinc ions. Miller et al. [17] later found that TiO$_2$ NPs exposed to low levels UVR, in simulated sunlight, caused toxic effects on marine plankton. The growth of marine phytoplankton was suppressed to varying extents for the different species studied and depended on the TiO$_2$ concentrations used, which ranged from 1 to 7 mg/L. In the absence of UVR, the growth rates of different marine plankton species were typically not affected by exposure to TiO$_2$ NPs, apart from *I. galbana,* when exposed to TiO$_2$ at the highest concentration of 7 mg/L used in the study.

Suman et al. [18] found that the cell viability of the marine algae species, *Chlorella vulgaris* (*C. vulgaris*), decreased with increasing ZnO NP concentration and duration of exposure. ZnO NPs were found to induce significant oxidative stress in *C. vulgaris* cells. Oxidative stress was assessed by measuring the effect of ZnO NPs on the activity of different enzymes. The levels of superoxide dismutase (an enzymatic antioxidant) and reduced glutathione (a nonenzymatic antioxidant) were found to decrease with increasing ZnO NP dose, suggesting that enzyme activity was activated at low ZnO NP dose but inhibited at higher doses of ZnO NPs and indicating that ZnO NPs increased oxidative stress. The increasing lipid peroxidation levels with ZnO NP dose, further indicated oxidative stress due to ZnO NPs. The levels of LDH, which increased with ZnO NP dose and/or longer exposure durations, indicated cell membrane damage on exposure to ZnO NPs, which may cause cell death. However, the high concentrations of ZnO NP dispersions selected for this study, varying from 50 to 300 mg/L, are not typical of the ZnO concentrations that would occur in marine or other aquatic environments.

The effect of TiO_2 NPs (Aeroxide P25) on two species of freshwater green microalgae, namely *Chlorella* and *Scenedesmus*, was investigated in a study by Roy et al. [19]. The three TiO_2 concentrations used in this study, 0.12, 1.25, and 12.52 μM, were selected to simulate levels of NPs in surface water. The colloidal stability and sedimentation tendency of NPs affected their toxic effects on the living cells. The hydrodynamic size of TiO_2 NPs was found to increase but remained in the nanoscale. Significant differences in the toxic effects caused by *Chlorella* and *Scenedesmus* were only observed at the highest TiO_2 NP concentration (12.52 μM) under UVA radiation and visible light conditions. TiO_2 NPs caused greater cell damage to the single-cell *Chlorella* species, while *Scenedesmus* was less susceptible to the toxicity of TiO_2 due to its tendency to form colonies. The oxidative stress induced in the cells of the two freshwater algae species exposed to TiO_2 NPs, as assessed from the activities of superoxide dismutase, glutathione reductase and catalase and the lipid peroxidation level, correlated well to the observed cytotoxic effects.

Carvalhais et al. [20] observed only mild effects in a study conducted to investigate the toxicity of TiO_2 NPs and oxybenzone, individually and in a mixture, to a marine flatfish (*Scophthalmus maximus*). In this study, the UV filters were intraperitoneally injected into the fish, and the oxidative stress, neurotoxicity, and metabolic profile were assessed after three and seven days.

Bivalve molluscs are found to be particularly susceptible to the toxic effects of inorganic UV filter NPs due to their unique internal mechanisms, such as phago-cytosis and endocytosis that are essential for cellular immunity and intracellular digestion and processes [21].

Canesi et al. [21] investigated the effects of commercial TiO_2, SiO_2 and C_{60} fullerene NPs in vitro on *Mytilus* hemocytes. The three types of NPs studied did not affect the stability of lysosomal membranes in assays, suggesting a lack of significant cytotoxic effects on hemocytes. Still, they did induce immune and inflammatory responses in cells, such as lysozyme release and the generation of nitric oxide and ROS, to different extents depending on the NP type and concentration.

D'Agata et al. [22] investigated the effects of bulk TiO_2 and TiO_2 NPs (as-obtained and aged) on *Mytilus galloprovincialis*, a species of bivalve. All forms of TiO_2 accumulated preferentially in the digestive glands, over in the gills, with both forms of TiO_2 NPs accumulating significantly (>six-fold) more than bulk TiO_2 in the digestive gland. However, greater tissue damage was observed in the digestive gland and gills exposed to bulk TiO_2, suggesting its higher toxicity *Mytilus galloprovincialis* than TiO_2 NPs. The three forms of TiO_2 were found to increase DNA damage in haemocytes when compared to the control, but the extent of damage was similar for all three forms of TiO_2.

Commercial coated and uncoated TiO_2 NPs, dispersed in either cosmetic oil or water, depending on the hydrophilicity or hydrophobicity of the NPs, did not adversely affect the immunity or development of the sea urchin, *Paracentrotus lividus*, at concentrations similar to the NP levels expected in seawater near the shore during summer, in studies by Catalano et al. [23]. The shape of hydrophilic TiO_2 NPs influenced their interaction with sea urchin embryos, with rod-shaped particles slightly impacting their development more than spherical particles. The coatings of hydrophobic TiO_2 NPs, and their interactions with the cosmetic oil medium influenced their effects on urchin embryos.

Buffet et al. [24] employed isotopically labelled ZnO NPs to investigate the fate of ZnO NPs in sediment and their effects on two estuarine intra-sedimentary invertebrates, the clam *Scrobiculariaplana* and ragworm *Nereis diversicolor*, to differentiate Zn arising from ZnO NPs exposure to Zn naturally occurring in sediment. The concentration of ^{67}Zn in the sediment progressively decreased with increasing sediment depth, from 2.59 mg/kg at 1 cm depth to 0.31 mg/kg at 5 cm depth, in sediment exposed to 3 mg/kg levels of ZnO NPs dispersed in diethylene glycol. ^{67}Zn bioaccumulated in both the clam and ragworm species studied to a similar extent. However, the effects on clam and ragworm due to ^{67}Zn could not be established as similar effects were observed in diethylene glycol (DEG) without ZnO NPs.

Widely used inorganic and organic UV filters that persist and co-exist in the environment have been detected in soil, water bodies, and sediments [25]. Sun et al. [25] investigated the effect of co-exposure to benzophenone-3 (BP3) and TiO_2 NP on the development of zebrafish. Zebrafish embryos were exposed to a mixture of TiO_2 NPs and BP3 and TiO_2 NPs and BP3 individually from 6 to 24 h after fertilization, and samples were taken at 48, 60 and 72h after fertilization for assessment of different toxic effects. Based on studies in the literature, environmentally relevant concentrations of TiO_2 NPs and BP3 of 100 and 10 ug/L were selected for this study, respectively. This study found that exposure to TiO_2 NPs and BP3 individually and in combination resulted in different neurotoxic effects in Zebrafish embryos. Certain toxicological effects such as cell apoptosis, axonal growth inhibition, and changes to motor behaviours of larvae were found to be higher in groups co-exposed to BP3 and TiO_2 NPs than groups exposed individually to TiO_2 NPs or BP3, which was correlated to the higher ROS levels observed for the co-exposed group than the other two groups.

The studies from the literature presented above have indicated that TiO_2 and ZnO particles have the propensity to cause different effects on plant and animal cells and organisms, both in the absence of UVR or under UV irradiation. The observed toxic effects are mainly attributed to ROS generated by the photocatalytic metal oxide particles and to free metal ions arising from the dissolution of metal oxide particles in water or soil, especially in the case of Zn^{2+}. However, it should be noted that the concentrations of TiO_2 and ZnO selected for toxicology studies are often higher than the levels that typically occur in the environment, sometimes by several orders of magnitude, and experiments are carried out under standard laboratory conditions where all risk factors are not considered. As for the human safety studies, the tests used for ecotoxicology studies are not standardized methods for toxicity assessments [10]. Furthermore, assessment of the potential effects of inorganic UV filters in the environment is a complex process which requires an understanding of the behaviour and fate of inorganic UV filters, which are affected by both the physiochemical properties of UV filter particles and different environmental conditions such as ionic strength, pH, and the presence and type of natural organic matter, and which requires both experimental studies, incorporating analytical techniques, and modelling methods [5].

References

1. Lozano C, Givens J et al (2020) Bioaccumulation and toxicological effects of UV filters on marine species. In: Sunscreens in coastal ecosystems: occurrence, behavior, effect and risk. The handbook of environmental chemistry, vol 94. Springer, Berlin/Heidelberg, Germany, pp 85–130
2. Chatzigianni M, Pavlou P et al (2022) Environmental impacts due to the use of sunscreen products: a mini-review. Ecotoxicology 31(9):1331–1345
3. Lebaron P (2022) UV filters and their impact on marine life: state of the science, data gaps, and next steps. J Eur Acad Dermatol Venereol 36(S6):22–28
4. Yuan S, Huang et al (2022) Environmental fate and toxicity of sunscreen-derived inorganic ultraviolet filters in aquatic environments: a review. Nanomaterials (Basel) 12(4):699
5. Heilgeist S, Sekine R et al (2021) Finding nano: challenges involved in monitoring the presence and fate of engineered titanium dioxide nanoparticles in aquatic environments. Water 13(5):734
6. Schneider SL, Lim HW (2019) A review of inorganic UV filters zinc oxide and titanium dioxide. Photodermatol Photoimmunol Photomed 35(6):442–446
7. Ma H, Kabengi NJ et al (2011) Comparative phototoxicity of nanoparticulate and bulk ZnO to a free-living nematode Caenorhabditis elegans: the importance of illumination mode and primary particle size. Environ Pollut 159(6):1473–1480
8. Egambaram OP, Pillai SK et al (2020) Materials science challenges in skin UV protection: a review. Photochem Photobiol 96(4):779–797
9. Cuccaro A, Oliva M et al (2022) Biochemical response of Ficopomatus enigmaticus adults after exposure to organic and inorganic UV filters. Mar Pollut Bull 178:113601
10. Freitas Neto LL, Espósito BP (2023) Toxicity of zinc oxide to scleractinian corals and zooxanthellae: a brief review. Quim Nova 46(3):266–272
11. Miller IB, Pawlowski S et al (2021) Toxic effects of UV filters from sunscreens on coral reefs revisited: regulatory aspects for "reef safe" products. Environ Sci Eur 33:74
12. Reinosa JJ, Docio CMÁ et al (2018) Hierarchical nano ZnO-micro TiO_2 composites: high UV protection yield lowering photodegradation in sunscreens. Ceram Int 44(3):2827–2834
13. Pawlowski S, Moeller M et al (2021) UV filters used in sunscreens—a lack in current coral protection? Integr Environ Assess Manag 17(5):926–939

14. Corinaldesi C, Marcellini C et al (2018) Impact of inorganic UV filters contained in sunscreen products on tropical stony corals (*Acropora spp.*). Sci Total Environ 637–638:1279–1285

15. Blaise C, Gagné F et al (2008) Ecotoxicity of selected nano-materials to aquatic organisms. Environ Toxicol 23(5):591–598

16. Miller RJ, Lenihan HS et al (2010) Impacts of metal oxide nanoparticles on marine phytoplankton. Environ Sci Technol 44(19):7329–7334

17. Miller RJ, Bennett S et al (2012) TiO_2 nanoparticles are phototoxic to marine phytoplankton. PLoS ONE 7(1):e30321

18. Suman TY, Rajasree SRR et al (2015) Evaluation of zinc oxide nanoparticles toxicity on marine algae chlorella vulgaris through flow cytometric, cytotoxicity and oxidative stress analysis. Ecotoxicol Environ Saf 113:23–30

19. Roy R, Parashar A et al (2016) Differential effects of P25 TiO_2 nanoparticles on freshwater green microalgae: Chlorella and Scenedesmus species. Aquat Toxicol 176:161–171

20. Carvalhais A, Pereira B (2021) Mild effects of sunscreen agents on a marine flatfish: oxidative stress, energetic profiles, neurotoxicity and behaviour in response to titanium dioxide nanoparticles and oxybenzone. Int J Mol Sci 22(4):1567

21. Canesi L, Ciacci C et al (2010) In vitro effects of suspensions of selected nanoparticles (C_{60} fullerene, TiO_2, SiO_2) on Mytilus hemocytes. Aquat Toxicol 96(2):151–158

22. D'Agata A, Fasulo S et al (2014) Enhanced toxicity of 'bulk' titanium dioxide compared to 'fresh' and 'aged' nano-TiO_2 in marine mussels (Mytilus galloprovincialis). Nanotoxicology 8(5):549–558

23. Catalano R, Labille J et al (2020) Safety evaluation of TiO_2 nanoparticle-based sunscreen UV filters on the development and the immunological state of the sea urchin Paracentrotus lividus. Nanomaterials (Basel) 10(11):2102

24. Buffet PE, Amiard-Triquet C et al (2012) Fate of isotopically labeled zinc oxide nanoparticles in sediment and effects on two endobenthic species, the clam Scrobicularia plana and the ragworm Hediste diversicolor. Ecotoxicol Environ Saf 84:191–198

25. Sun X, Yang Q et al (2023) Environmentally relevant concentrations of organic (benzophenone-3) and inorganic (titanium dioxide nanoparticles) UV filters co-exposure induced neurodevelopmental toxicity in zebrafish. Ecotoxicol Environ Saf 249:114343

Chapter 7
Current Regulations Related to the Use of Inorganic UV Filters in Sunscreens

Currently, ZnO and TiO_2 are the only two inorganic filters approved by international regulation bodies such as the EC and USFDA for use in sunscreens [1–5]. In the USA, sunscreens are classified as over-the-counter (OTC) drugs by the FDA and by itself regulated to a greater extent than in other parts of the world, such as in Europe, where sunscreens are classified as cosmetics. In other countries, sunscreens can be classified as either therapeutic drugs or cosmetics, depending on their properties, as in Australia, or as "quasi-drugs," as in Japan [5, 6]. The USFDA has supported a proposal that sunscreen products containing ZnO, TiO_2, or both in combination as active ingredients, at concentrations of 25% or less, are generally recognized as safe and effective (GRASE) under the Sunscreen Innovation Act on condition that all the additional requirements stipulated in the final sunscreen order are satisfied. This proposition was made after a review of available scientific literature and submissions to the USFDA's Adverse Event Reporting System and the sunscreen monograph docket related to the safety of ZnO and TiO_2 in sunscreens. The FDA has acknowledged the use of TiO_2 and ZnO NPs in sunscreen products but is not presently distinguishing their nanoparticle form from their other forms nor proposing conditions for their use under the sunscreen monograph, citing large data gaps and the need for further investigation into their safety for use [4].

Similarly, the Scientific Committee on Consumer Safety (SCCS), which advises the EC, has concluded that TiO_2 and ZnO, including their nanoparticle forms, in sunscreens are safe for use at levels up to 25% and do not pose any significant risk to human health when applied on healthy, undamaged or sunburnt skin [7–10]. This recommendation applies to TiO_2 NPs "with a median particle size based on the number size distribution of 30–100 nm or larger" and ZnO NPs with "a median diameter of the particle number size distribution above 30 nm" [11, 12]. However, the SCCS has discouraged the use of sunscreen products containing TiO_2 NPs or ZnO NPs that could result in lung exposure to NPs through inhalation, such as aerosol and powder-based formulations, due to evidence of their harmful effects, such as causing inflammation in lungs and risk of causing cancer from human and animal studies [7–9]. TiO_2 has been regarded as a possible carcinogen to humans, applicable mainly to TiO_2 exposure via the inhalation route [13]. Furthermore, the SCCS later reviewed their position on TiO_2 NPs in sunscreen and dissuaded the use of highly photocatalytic TiO_2 NPs, anatase, within cosmetic and sunscreen products [12, 14]. Even though TiO_2 NPs, at concentrations up to 25%, are permitted, the amount of anatase-TiO_2 NPs allowed was reduced to 5% of the total contained TiO_2 NPs, which equates to a concentration of 1.25% in the final product [12]. Therefore, the permitted use of TiO_2 and ZnO in cosmetics and sunscreens is tentative at best, being under constant review and subject to new information related to their safety.

As described before, for effective photoprotection of skin from the harmful effects of UVR, sunscreen products must offer broad-spectrum protection across the UVA/B region [15, 16]. The USFDA requires that a sunscreen product pass the broad-spectrum test and meet certain requirements to be labelled as broad-spectrum [17]. In addition, the SPF value, as determined from the stipulated SPF test procedure, must be indicated together with the broad-spectrum label [17]. Furthermore, as the SPF value does not indicate protection against the longer-wave UVAII radiation, it is imperative that the UVA protective factor (UVA-PF) of a sunscreen is also determined and taken into account [15, 18]. As a result, the USFDA has recommended that the UVA-PF should be at the minimum one-third of the sun protection factor (SPF) and that the critical wavelength, the wavelength at which 90% of the area under the spectral absorbance curve occurs, for the range 290–400 nm, be 370 nm or longer for a sunscreen product to be categorized as broad-spectrum [1, 13, 15, 17, 18]. The SCCS has similar guidelines to the FDA with regard to the labelling and testing of sunscreen and products in the EU [5]. Products can be classified as sunscreen products and are stamped with the UVA seal if the following requirements are met: the SPF value is 6 or higher, the UVA-PF is 1/3 of the SPF, and the critical wavelength is 370 nm or more [5].

The current regulations related to the use of ZnO and TiO_2 for photoprotection are subject to review based on new information on their potential effects in the human body and impacts in the environment which will largely dictate the future direction of their use in sunscreens. Consequently, there has been an increased investigation into the modification of ZnO and TiO_2 UV filters in order to ameliorate undesirable features for regulatory compliance, and alternative metal oxides and inorganic materials for use in UV filters, which will be expanded upon in the sections that follow.

References

1. Egambaram OP, Pillai SK et al (2020) Materials science challenges in skin UV protection: a review. Photochem Photobiol 96(4):779–797
2. Adler BL, DeLeo VA (2020) Sunscreen safety: a review of recent studies on humans and the environment. Curr Dermatol Rep 9:1–9
3. Narla S, Lim HW (2020) Sunscreen: FDA regulation, and environmental and health impact. Photochem Photobiol Sci 19(1):66–70

4. U.S. Food and Drug Administration (2021) Proposed order (OTC000008): amending over-the-counter (OTC) monograph M020: sunscreen drug products for OTC human use. https://dps-admin.fda.gov/omuf/omuf/sites/omuf/files/primary-documents/2022-09/Proposed%20Admi nistrative%20Order%20OTC000008_Amending%20M020_Sunscreen_Signed24Sept2021. pdf. Accessed 22 Feb 2024

5. Pirotta G (2020) Sunscreen regulation in the world. In: Sunscreens in coastal ecosystems: occurrence, behavior, effect and risk. The handbook of environmental chemistry, vol 94. Springer, Berlin/Heidelberg, Germany, pp 15–35

6. Mancuso JB, Maruthi R et al (2017) Sunscreens: an update. Am J Clin Dermatol 18(5):643–650

7. Schilling K, Bradford B et al (2010) Human safety review of "nano" titanium dioxide and zinc oxide. Photochem Photobiol Sci 9(4):495–509

8. Dréno B, Alexis A et al (2019) Safety of titanium dioxide nanoparticles in cosmetics. J Eur Acad Dermatol Venereol 33(S7):34–46

9. Uter W, Gonçalo et al (2014) Coupled exposure to ingredients of cosmetic products: III. Ultraviolet filters. Contact Dermatitis 71(3):162–169

10. Henkler F, Tralau T et al (2012) Risk assessment of nanomaterials in cosmetics: a European Union perspective. Arch Toxicol 86(11):1641–1646

11. SCCS (Scientific Committee on Consumer Safety) (2014) ADDENDUM to the opinion SCCS/1489/12 on Zinc oxide (nano form), 23 July 2013. Revision of 22 April 2014. http://ec.eur opa.eu/health/scientific_committees/consumer_safety/docs/sccs_o_137.pdf. Accessed 22 Feb 2024

12. Scientific Committee on Consumer Safety, Chaudhry Q (2015) Opinion of the scientific committee on consumer safety (SCCS)—revision of the opinion on the safety of the use of titanium dioxide, nano form, in cosmetic products. Regul Toxicol Pharmacol 73(2):669–670

13. Smijs TG, Pavel S et al (2011) Titanium dioxide and zinc oxide NPs in sunscreens: focus on their safety and effectiveness. Nanotechnol Sci Appl 4:95–112

14. Solaiman SM, Algie J et al (2019) Nano–sunscreens—a double–edged sword in protecting consumers from harm: viewing Australian regulatory policies through the lenses of the European Union. Crit Rev Toxicol 49(2):122–139

15. Serpone N (2021) Sunscreens and their usefulness: have we made any progress in the last two decades? Photochem Photobiol Sci 20(2):189–244

16. Diffey B (2009) Sunscreens: expectation and realization. Photodermatol Photoimmunol Photomed 25(5):233–236

17. Over-the-counter sunscreen drug products; required labelling based on effectiveness testing. Electronic Code of Federal Regulations Title 21, Section 201.327 (21CFR201.327). https://www.ecfr.gov/current/title-21/chapter-I/subchapter-C/part-201. Accessed 22 Feb 2024

18. Diffey BL, Tanner PR et al (2000) In vitro assessment of the broad-spectrum ultraviolet protection of sunscreen products. J Am Acad Dermatol 43(6):1024–1035

Chapter 8
Modification of ZnO and TiO$_2$ UV filters

Researchers have investigated different strategies for improving the properties of inorganic UV filters used in sunscreens in order to mitigate their undesirable features, such as intrinsic photocatalytic activity, while still maintaining their UV attenuation properties [1–8]. One method used for reducing the photoreactivity of TiO$_2$ and ZnO has been their encapsulation with organic substances such as hexadecylamine, oleic acid and polymers and inorganic coatings such as silica, alumina, dimethicone, methicone, aluminium hydroxide, aluminium oxide, and polymethylacrylic acid, as the surface properties of particles affect their photoreactivity [1–6]. Coatings impede photocatalytic reactions at the particle surface by inhibiting direct interactions between metal oxide particles and skin cells or the formulation medium and by scavenging free radicals or preventing their formation [1, 3, 4, 6]. However, these coatings do not completely diminish the photocatalytic effects of ZnO and TiO$_2$ NPs and have even increased activity in some cases [1, 3, 4, 6].

Mitchnick et al. [9] investigated the photoreactivity of microfine ZnO towards organic UV filters in combination sunscreens, with a composition of 7.5% organic filter/10% microfine ZnO. It was found that the two filters studied, OMC and avobenzone, were not photodegraded by ZnO after irradiation at UVR doses of 10, 20, and 30 J/cm^2. The microfine ZnO used for this study was a mixture of noncoated and dimethicone-coated particles in equal amounts. The effect of particle coating on the photoreactivity of microfine ZnO and microfine TiO$_2$ towards isopropanol oxidation was also investigated in this study. Uncoated microfine TiO$_2$ showed higher photoreactivity than uncoated microfine ZnO. The coating of ZnO and TiO$_2$ particles with dimethicone and silica significantly decreased the photoreactivity towards the oxidation of isopropanol. The sunscreen containing 15% uncoated microfine ZnO (Z-Cote) was considered to be photostable based on the similarity of the absorption curves obtained before and after irradiation of the sunscreen by a solar simulator.

Carlotti et al. [10] investigated the efficacy of the different coatings of commercially available TiO$_2$ NPs to attenuate the inherent photoreactivity of TiO$_2$. The photocatalytic activity of TiO$_2$ was evaluated by measuring free radical generation

N. H. Kera et al., *Inorganic Ultraviolet Filters in Sunscreen Products*, SpringerBriefs in Materials, https://doi.org/10.1007/978-3-031-64114-5_8

using ESR spectroscopy and by quantifying lipoperoxidation of linoleic acid and porcine ear skin. Covasil S1 (trymethoxycaprylylsilane-coated, 80% anatase, 20% rutile) and T-Lite SF (Al(OH)$_3$ coated, rutile) showed significant activity towards the peroxidation of linoleic acid, similar to that of uncoated TiO$_2$ (Aeroxide P25). T-Lite SF-S (Al(OH)$_3$ and SiO$_2$ coated, rutile), Maxlight F-TS20 (SiO$_2$-coated, rutile), and TEGO Sun TS Plus (SiO$_2$ and trymethoxy-caprylylsilane-coated, 80% anatase, 20% rutile) showed low photocatalytic activity, on par with that of the control. The findings of this study suggested that silica-based coatings were stable and effective in reducing the photocatalytic activity of TiO$_2$ NPs.

Yin et al. [11] compared the photoreactivity, cytotoxicity and genotoxicity of pristine ZnO NPs to ZnO NPs coated with poly(methacrylic acid) (PMAA), oleic acid (OA), and adsorbed cell culture medium components. Cytotoxicity was deduced from the effect of uncoated and coated ZnO, at varying concentrations, on human lymphoblastoid (WIL2-NS) cells after different exposure durations. Cell viability increased on exposure to low concentrations of uncoated ZnO NPs for short durations, such as 2 mg/L (4 and 24 h) and 10 mg/L (4 h). Still, it was decreased by higher ZnO concentrations and at longer exposure durations. For coated ZnO NPs, cellular activity was found to decrease with an increase in NP concentration and exposure duration. The cell viabilities obtained for the three coated ZnO samples were typically higher than that of uncoated ZnO, indicating their lower cytotoxicity in the order PMMA $\approx$ cell culture medium < OA (< uncoated ZnO). The generation of ROS by the coated and uncoated ZnO samples in contact with WIL2-NS cells was determined using a fluorescence assay that measures hydroxyl radicals. The ROS generated with coated ZnO NPs was in the order: cell culture medium < PMMA coated < OA coated < uncoated ZnO and appeared to be inversely related to their cytotoxicity. However, PMMA-coated ZnO exhibited significantly higher genotoxic effects than uncoated ZnO and the other two coated samples.

Bartoszewska et al. [12] modified TiO$_2$ NPs by coating with silver and silica. TiO$_2$/SiO$_2$ and TiO$_2$/Ag nanostructures were prepared via different two-step methods, as shown in Fig. 8.1a, through careful control of experimental conditions, and their properties compared to that of TiO$_2$ NPs, also prepared in this study. The UV-visible spectra obtained for TiO$_2$ NPs and TiO$_2$/Ag nanostructures (Fig. 8.1a) showed similar absorption maxima, while for TiO$_2$/SiO$_2$ nanostructures, the absorption maximum shifted towards the UVA region. Water contact angle measurements indicated that the contact angle was 74.8° for TiO$_2$/Ag nanostructures, making it the least hydrophilic, when compared to TiO$_2$ NPs and TiO$_2$/SiO$_2$ nanostructures, for which contact angles of 0° and 4.6° were determined, respectively. This suggested the lower susceptibility of TiO$_2$/Ag nanostructures to being washed off the skin when used as a UV filter in sunscreens. An oil/water emulsion containing TiO$_2$/Ag nanostructures had slightly longer sun protection times than emulsions containing TiO$_2$ NPs and TiO$_2$/SiO$_2$ nanostructures and the blank emulsion, with measurements being made before and at 20 min after emulsion application on pig skin and again after 25 min of washing. The authors suggested that the longer sun protection times obtained for emulsions containing TiO$_2$/Ag nanostructures was due to its suitable particle size (~50 nm) for use as a UV filter and the incorporation of silver, which may have enhanced the SPF.

However, SPF values of sunscreen emulsions were not provided in this study. The authors also did not report on the effect of Ag and SiO$_2$ coatings on the photocatalytic activity of TiO$_2$.

Choi et al. [13] obtained SiO$_2$-TiO$_2$-Tannin (TA) hybrid particles, as depicted in Fig. 8.1b, for application as UV-to-blue light filters by firstly preparing SiO$_2$-TiO$_2$ particles via a sol–gel method wherein TiO$_2$ NPs were deposited on the surface on SiO$_2$ and then dispersing the SiO$_2$-TiO$_2$ particles in a tannic acid solution. UV– visible absorption spectra showed that SiO$_2$-TiO$_2$ particles and SiO$_2$-TiO$_2$-Tannin particles both absorbed strongly in the UV region and had similar absorption profiles. The UV–visible absorption spectrum for SiO$_2$-TiO$_2$-Tannin particles (Fig. 8.1b) also showed an additional broadband in the visible light region, indicating the generation of a new photoexcitation route. The bandgap energy values obtained from Tauc plots were 3.2 eV and 1.90 eV for SiO$_2$-TiO$_2$ and SiO$_2$-TiO$_2$-TA, respectively. The levels of hydroxyl radicals and superoxide anions generated on UV irradiation of SiO$_2$-TiO$_2$- TA particles were significantly lower than that observed for SiO$_2$-TiO$_2$ particles and TiO$_2$ NPs. The authors suggested that galloyl moieties of the TA layer, with two hydroxyl groups in ortho positions, suppressed ROS generation by shifting between its three forms: radical, galloyl, and quinone, as shown in Fig. 8.1b. Suppression of ROS generation by the TA layer of SiO$_2$-TiO$_2$-TA particles was evident in dye photodegradation and cytotoxicity studies. Rhodamine B photodegradation by SiO$_2$- TiO$_2$-TA particles was insignificant compared to TiO$_2$ NPs and SiO$_2$-TiO$_2$ particles, which degraded RB by 100% and 33%, respectively, after 1 h. Cell viability was higher in NIH-3 T3 cells treated with SiO$_2$-TiO$_2$-TA particles (75%) than untreated cells (59%) under UVA irradiation, suggesting a protective effect.

In contrast, TiO$_2$ NPs and SiO$_2$-TiO$_2$ particles with cell viabilities lower than and about the same as that of the untreated cells, respectively, did not protect cells from damage due to UVA exposure due to photogenerated ROS harming cells. Furthermore, under blue light irradiation, higher cell viabilities were obtained for cells treated with SiO$_2$-TiO$_2$-TA particles than those treated with TiO$_2$ NPs and SiO$_2$-TiO$_2$ particles, which had cell viabilities similar to the blank (untreated cells). Being microsized, the SiO$_2$-TiO$_2$-TA particles could not penetrate the cell membranes of fibroblast cells, unlike TiO$_2$ NPs, which entered cells and were found in the cytoplasm.

The encapsulation of inorganic UV filters by different synthetic polymeric materials, such as polystyrene, poly(methyl methacrylate), and poly(methyl methacrylate-block-acrylic acid), has also been investigated [13, 14]. However, these polymer coatings were susceptible to decomposition in reactions involving ROS generated on irradiation of metal oxide-based UV filters [13, 14]. The use of microplastics in cosmetics is currently disfavoured and even banned in some cases due to their status as ubiquitous pollutants in the environment [13, 14]. As such, investigation of different biopolymers for encapsulation of inorganic UV filters has gained traction due to their attractive properties such as abundance in nature, nontoxicity, biocompatibility, and biodegradability, which render biopolymers ideal for use in skin products [15, 16]. Morlando et al. [15] reported on chitosan-encapsulated TiO$_2$ NPs obtained by using a spray-drying method. The diffuse reflectance spectra for the chitosan/TiO$_2$ nanocomposites showed a slight red-shift of the primary UV absorption bands when

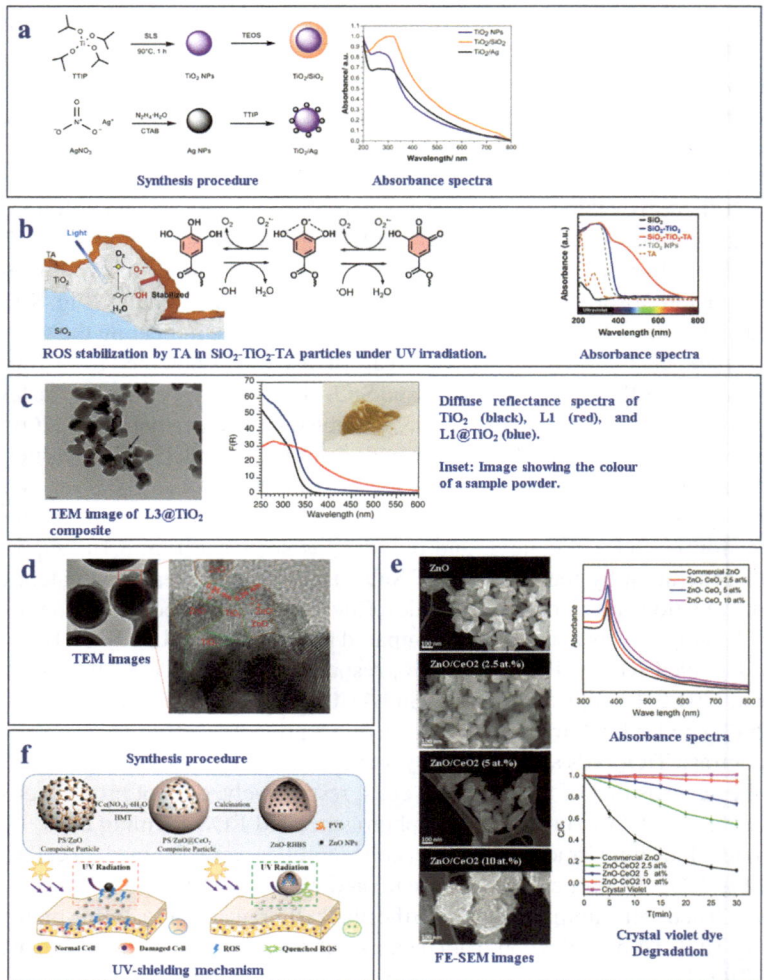

Fig. 8.1 Modification of ZnO and TiO₂ particles by different methods for application as UV filters: **a** Scheme showing synthesis procedure and absorbance spectra for TiO₂/SiO₂ and TiO₂/Ag nanostructures. Reproduced with permission from [12]. Copyright 2023, MDPI. **b** Absorbance spectra of SiO₂-TiO₂-Tannin hybrid particles and scheme showing ROS stabilization by TA layer of particles under UV irradiation [13]. **c** Diffuse reflectance spectra of Lignin@TiO₂ composite and typical TEM image [16]. **d** TEM images of TiO₂@ZnO porous hollow multishelled microspheres [28]. **e** SEM images and absorbance spectra of ZnO/CeO₂ nanocomposites and results of dye photodegradation studies. Reproduced with permission from [30]. Copyright 2020, Springer. **f** Schemes showing synthesis procedure and UV shielding mechanism of ZnO/CeO₂ hollow microcapsules [33]. **b, f** Reproduced with permission from Elsevier Science Ltd. **c, d** Reproduced with permission from American Chemical Society

compared to that of TiO_2 NPs and contained other smaller bands in the visible light region due to absorption by the chitosan encapsulation. The photocatalytic activity of chitosan/TiO_2 nanocomposites was significantly lower than TiO_2 NPs towards the degradation of crystal violet dye over 2 h, with the degradation efficiency decreasing with increasing chitosan content. The decrease in dye degradation was attributed to the chitosan coating hindering the movement of photogenerated charge carriers to the particle surface, thus inhibiting the generation of ROS and free radicals and the decreased adsorption of crystal violet dye due to a decrease in TiO_2 surface area as a result of the agglomeration of chitosan-coated TiO_2 particles.

Morsella et al. [16] prepared lignin@TiO_2 composites by encapsulation of TiO_2 NPs with different kinds of lignin, ranging from aqueous-soluble lignins to organosoluble lignins. The lignin@TiO_2 composites were prepared by UVA irradiation of a mixture of TiO_2 and lignin solution (in an organic or aqueous solvent). This method exploited the photocatalytic activity of TiO_2 as the polymeric species of lignin were crosslinked by the ROS generated on UVA irradiation and formed a shell around TiO_2 particles, as can be seen in the TEM image for L3@TiO_2 composite in Fig. 8.1c. The intensity of absorption differed for the different lignin@TiO_2 composites, with the composite containing the organosoluble, kraft lignin (L1) showing the best UV attenuation. The diffuse reflectance spectra obtained (Fig. 8.1c) showed that the UV absorption profile of TiO_2 was slightly extended into the visible region by the lignin coating (L1). Higher lignin loadings in the composites were obtained with the organosoluble lignins, which also exhibited higher stability than the composites containing water-soluble lignins, measured in terms of the lignin released or degraded on UV irradiation of aqueous solutions of lignin@TiO_2 composites for 2 h. The lignin@TiO_2 composites exhibited lower photocatalytic activity than TiO_2 towards the oxidation of 2-Propanol under UVA–UVB irradiation. The enzymatic activity of alkaline phosphatase(ALP) was decreased on contact with TiO_2 in the dark and, to a greater extent, under UV light. In contrast, contact with lignin@TiO_2 composites did not inhibit ALP enzymatic activity under dark conditions and resulted in higher enzymatic activity than TiO_2 under UVA irradiation. The photodegradation of avobenzone in contact with varying amounts of lignin@TiO_2 nanocomposites was investigated under UVA–UVB irradiation for 2 and 4 h periods to assess the compatibility of the composites with other active ingredients in sunscreens. Avobenzone is a commonly used UVA filter that rapidly degrades on exposure to UV in sunlight. TiO_2 was found to protect avobenzone from photodegradation when present at lower particle doses of 0.01 and 0.03 wt% but offered no protection at the higher dose of 0.06 wt%, possibly due to the high levels of ROS generated. The presence of lignin@TiO_2 nanocomposites reduced the photodegradation of avobenzone at all particle doses studied and to a greater extent than TiO_2. The photocatalytic activity of TiO_2 was reduced by coating with lignin, which inhibited free radical reactions due to its antioxidant properties, possibly by scavenging the ROS generated by UV-irradiation of TiO_2.

A review article by Kim et al. [17] presented further examples of the use of lignin and other biomass-derived materials for improving sunscreens containing TiO_2 or

ZnO [16]. In addition to being used as coating agents for TiO$_2$ and ZnO NPs, biomass-derived materials have also been used for the synthesis of NPs, as capping and stabilizing agents, as additives in formulations and incorporated into composites containing inorganic UV filters.

Loto et al. [18] prepared a series of TiO$_2$ @plant extract nanocomposites wherein TiO$_2$ was coated with aqueous plant extracts obtained from the branches, leaves, fruits, and bark organs of the *Sarcomphalus mistol* and *Schinopsis lorentzii* tree species, native to the Chaco region of Argentina. The diffuse reflectance spectra showed that TiO$_2$ @plant extract nanocomposites showed higher absorption in the UVA and visible region than TiO$_2$. In contrast, the optical density spectrum of TiO$_2$ @PE suspensions showed more intense bands than that of TiO$_2$. The TiO$_2$ @plant extract nanocomposites showed almost no photocatalytic activity towards the photodegradation of rhodamine B dye under UVA light, in contrast to TiO$_2$, which completely degraded the dye. The authors suggested that the different components of the plant extract coating contributed towards diminishing the photocatalytic activity of TiO$_2$ by scavenging radicals or reactive species formed on UV irradiation of TiO$_2$ by absorbing UV radiation before it reached the TiO$_2$ particles and by preventing the adsorption and reaction of dye molecules on the TiO$_2$ surface. The plant extract coating of a selected nanocomposite (TiO$_2$ @QC_bark) exhibited good stability, measured in terms of retained photoprotective and colloidal properties, in both seawater and freshwater media used for simulating environmental conditions. The in vitro determined SPF and UVA-PF values of aqueous suspensions of TiO$_2$ @PE were, on average, higher than that obtained for TiO$_2$.

Serpone et al. [19] reported on photo-inactivated TiO$_2$ samples with significantly decreased photoactivity obtained by surface modification of particles through a thermally assisted method. The photo-inactivated TiO$_2$ samples displayed reduced photocatalytic activity towards phenol photo-oxidation. They caused less damage to DNA plasmids, human cells, and yeast cells exposed to UVB/UVA in simulated solar light than the un-modified TiO$_2$ samples. Serpone et al. [4] further suggested a simple sunscreen with high photostability consisting of photo-inactivated TiO$_2$ particles in an aloe vera gel-base medium.

Abuçafy et al. [20] prepared transparent gel-based formulations, prepared with Carbopol 940, incorporating *p*-toluene sulfonic acid (PTSH) surface-modified TiO$_2$ NPs to prevent the undesirable white cast that results when TiO$_2$-containing sunscreen formulations are applied on the skin. The modification of TiO$_2$ NPs imparted a negative surface charge to the particles, which prevented their aggregation and increased their apparent transparency in sunscreens. The UV-attenuation profile of sunscreen formulations containing 10 and 15% of PTSH-modified TiO$_2$ NPs was like that of formulations containing commercial TiO$_2$ particles. Still, absorbance intensities were lower for the surface-modified TiO$_2$ NPs. Diffuse reflectance spectra in the UV-visible region showed that formulations containing PTSH-modified TiO$_2$ NPs exhibited significantly lower reflectance than formulations containing commercial TiO$_2$, indicating the higher transparency of PTSH-modified TiO$_2$ NPs. The gel formulations containing commercial TiO$_2$ had a higher viscosity than those containing PTSH-modified TiO$_2$ NPs due to the tendency of unmodified TiO$_2$ NPs to flocculate in the formulation medium.

Battistin et al. [21] functionalized the surface of nanometric and non-nanometric TiO$_2$ with Oxisol (dihydroxyphenyl benzimidazole carboxylic acid), which involved the coordination of mainly catechol groups and, to a lesser extent, carboxylic acid functionalities, with the surface of TiO$_2$. Surface functionalization improved the sedimentation rates and indicators of suspension stability of both nanometric and non-nanometric TiO$_2$, with the increase in stability being linked to steric repulsions. Photochemiluminescence assays showed that functionalized nanometric TiO$_2$ displayed antioxidant activity similar to the Oxisol reference but significantly higher than the functionalized non-nanometric TiO$_2$ and the nanometric/Oxisol and non-nanometric/Oxisol physical mixtures. The high antioxidant activity of functionalized nanometric TiO$_2$ was attributed to its higher surface area, greater amount of Oxisol functionalization through carboxyl group coordination, and lower oxidation potential of TiO$_2$ when compared to the functionalized non-nanometric TiO$_2$ and the mixture samples. Surface functionalized TiO$_2$ samples exhibited lower photocatalytic activity than unfunctionalized TiO$_2$, for both nanometric and non-nanometric TiO$_2$, in dye degradation studies carried out in the dark and under UV irradiation. Cytotoxicity studies indicated that the growth of 3T3 cells was not inhibited by nanometric and non-nanometric TiO$_2$ and their functionalized counterparts for concentrations varying from 10 to 100 μg/mL.

Metal ion doping of the crystal lattices of TiO$_2$ and ZnO or the deposition of metal/oxide particles on their surfaces are other techniques used to reduce their photoreactivity [1, 7]. Dao et al. [22] investigated the UV absorption properties and photoreactivity of Al, Fe, and Cu-doped ZnO NPs obtained in a simple coprecipitation method. The metal doping of ZnO NPs resulted in altered cell volumes, decreased crystallinity, and smaller particle sizes. The absorption spectra of undoped ZnO showed a broad band in the UV region, with a λ_{max} at 372 nm and no adsorption in the visible region. The UV-absorption capacity of ZnO was retained after doping with the different metal ions, although visible light absorption was increased in Fe- and Cu-doped ZnO NPs. The photocatalytic activity of doped ZnO NPs, assessed in terms of the photodegradation of methylene blue dye, was substantially lower than that of undoped ZnO. The reduced photoreactivity of the doped ZnO samples was linked to an increase in structural defects in the ZnO lattice because of metal ion doping, which could serve as recombination sites for electron–hole pairs, thus inhibiting photocatalytic activity. Similar findings were reported in a study by Le et al. [23], where Fe-doped ZnO samples with varying dopant concentrations were prepared by an oxalate coprecipitation. Al-doped TiO$_2$ NPs with the formula Al$_x$Ti$_{1-x}$O$_2$ (where $x = 0.03, 0.04$, and 0.05) were synthesized in a study by Ngoc et al. [24] via a sol–gel method. XRD analysis showed that the undoped TiO$_2$ NPs contained both rutile and anatase phases and that Al-doping of TiO$_2$ introduced a third phase, brookite. The decrease in cell volumes on doping also indicated Al-doping of TiO$_2$ as Al^{3+} has a smaller ionic radius than Ti^{4+}. The specific surface areas of Al-doped TiO$_2$ samples (69–79 m^2/g) were higher than that of undoped TiO$_2$ NPs (31 m^2/g). UV-visible absorption spectra showed that the undoped and Al-doped TiO$_2$ samples had similar absorption profiles, exhibiting strong absorbance in the UV region, with maxima at 210 and 340 nm, and weak absorbance in the visible region. The enhanced UV absorption of a sunscreen formulation containing the $Al_{0.04}Ti_{0.96}O_2$ powder, when compared to that containing TiO$_2$ NPs, was attributed to the higher anatase content and smaller particle size (and higher specific surface area) of the Al-doped sample. The rate constants determined for the photodegradation of methylene blue were lower for the Al-doped TiO$_2$ samples (falling within the 0.17–0.41 h^{-1} range) than the undoped TiO$_2$ NPs (1.14 h^{-1}), indicating that Al-doping decreased the photocatalytic performance of TiO$_2$ NPs. Differences in the colour of sunscreens before and after UV irradiation were more pronounced in those containing TiO$_2$ NPs than in Al-doped TiO$_2$ samples and attributed to the degradation of organic components in the sunscreens containing TiO$_2$ NPs due to their photocatalytic activity.

However, even though doping can enhance the properties of inorganic UV filters, potential metal dopants must be screened as doping has increased the photoreactivity of some materials, as in a study by Khairy et al. [25], where Cu and Zn doping of TiO$_2$ NPs decreased the band gap of TiO$_2$ and increased its catalytic efficiency towards the photodegradation of methyl orange dye and for the oxidation of glucose by KMnO$_4$.

Rampaul et al. [26] investigated the photoreactivity of different pristine and modified TiO$_2$ and ZnO UV filter NPs obtained from two sources, either from manufacturers in powder form or extracted from randomly selected commercial sunscreens. The UV filter NPs were extensively characterized to determine their crystal phase and type of modification, if present, such as doping or coating type. The results obtained suggested that the photocatalytic activity of UV filter NPs, assessed in terms of their capability to photodegrade methylene blue dye under UVA irradiation, was mainly dependent on the crystal phase of the UV filter NPs and modification method use and was not dependent on the surface area of NPs. The UV filter NPs that showed higher reactivity towards methylene blue degradation included both uncoated and coated TiO$_2$ of mixed anatase and rutile crystal form, coated and uncoated ZnO NPs, and both coated and uncoated mixed Rutile-TiO$_2$/ZnO powders samples. In comparison, coated and doped Rutile phase-TiO$_2$ samples typically displayed lower reactivity towards MB degradation. However, the findings from this study regarding the effect of coating/doping type and crystal phase on the photoreactivity of UV filters were not conclusive as comparisons could not be made due to lacking pertinent data such as the photoreactivity of pure and coated anatase and rutile forms. The photoreactivity of UV filters under UVB-irradiation was also not investigated here. The effects of the UV filter NP samples on cultured human skin cells correlated largely with the results of the methylene blue photodegradation studies. Both uncoated and coated TiO$_2$, of mixed anatase and rutile crystal form, were destructive towards cells, whereas coated and doped rutile phase-TiO$_2$ samples protected cells from harmful effects and death.

Reinosa et al. [27] synthesized a ZnO-TiO$_2$ composite composed of ZnO NPs deposited on TiO$_2$ microparticles by using a sol–gel method in an attempt to reduce the photoactivity of TiO$_2$ which is the main cause of the photodegradation of sunscreen formulations. XRD analysis confirmed the main crystalline phase to be the rutile form of TiO$_2$ and the presence of ZnO as a minor phase. The UV–vis absorption spectra obtained showed that the absorption maximum of the ZnO-TiO$_2$

composite occurred at 368 nm, at a wavelength in between that of ZnO NPs (λ_{max} = 370 nm) and TiO$_2$ microparticles (λ_{max} = 365 nm). Similarly, the band gap of the composite (3.08 eV) was in between that of ZnO NPs (3.05 eV) and TiO$_2$ microparticles (3.15 eV). SPF values of 3.77 and 4.45 were obtained for sunscreen formulations containing TiO$_2$ microparticles and ZnO-TiO$_2$ composite, respectively. The enhanced SPF value obtained for the latter was attributed to an increase in surface area and light scattering effects due to the anchoring of ZnO NPs on TiO$_2$ microparticles. After exposure to sunlight for 3 h, the SPF of the sunscreen formulation containing ZnO-TiO$_2$ composite decreased only slightly from 4.45 to 4.22 (~5%). Under the same conditions, the sunscreen formulation containing only TiO$_2$ microparticles showed discolouration, suggesting photodegradation of organic components, and a decrease of its SPF value from 3.77 to 2.35 (by 38%). The higher photostability of the sunscreen containing ZnO-TiO$_2$ composite was attributed to its hierarchical structure, which partially inhibits the formation of ROS and free radicals.

Parvin et al. [28] synthesized TiO$_2$@ZnO porous hollow multishelled microspheres (PHMMs) to circumvent the drawbacks of using NPs as UV filters. They compared its properties and performance to that of three samples: ZnO NPs, TiO$_2$ NPs, and a ZnO/TiO$_2$ NP (1:1) mixture. Transmission electron microscopy images of TiO$_2$@ZnO PHMMs are presented in Fig. 8.1d, indicated the formation of type II heterojunctions in regions where TiO$_2$ and ZnO NPs overlapped. It was determined from optical studies that TiO$_2$@ZnO PHMMs had a band gap of 4.33 eV and absorption maximum at 302 nm, compared to that of ZnO and TiO$_2$ NPs with band gap values of 3.34 and 3.23 eV and absorption maxima at 371 nm and 317 nm, respectively. TiO$_2$@ZnO PHMMs exhibited the lowest photocatalytic activity of the three samples studied, measured in terms of methylene blue photodegradation. High cell viabilities were maintained when human dermal fibroblasts were exposed to TiO$_2$@ZnO PHMMs for 24 h for all concentrations studied, ranging from 1 to 10 μg/mL.

In contrast, cell viabilities decreased with increased concentrations of ZnO NPs, TiO$_2$ NPs, and the ZnO/TiO$_2$ NP (1:1) mixture. It was suggested that the wide band gap of TiO$_2$@ZnO PHMMs and the heterojunctions within its structure resulted in decreased ROS formation and contributed towards the lower photocatalytic activity and less cytotoxic nature of TiO$_2$@ZnO PHMMs when compared to TiO$_2$ and ZnO NPs. The SPF of a sunscreen formulation incorporating TiO$_2$@ZnO PHMs was higher than that of formulations containing the other samples studied and that of a commercial sunscreen. However, the wider band gap of TiO$_2$@ZnO PHMMs could compromise its UVA absorption broad-spectrum protection, as indicated by the red-shift of absorption maxima compared to that of TiO$_2$ and ZnO.

To reduce the photocatalytic activity and cytotoxic effects of TiO$_2$ in sunscreens, Borrás et al. [29] prepared Y$_2$O$_3$ decorated TiO$_2$ (TiO$_2$@Y$_2$O$_3$) NPs via a hydrothermal method. The TiO$_2$@Y$_2$O$_3$ NPs showed higher UV absorption in the 280–350 nm region and less visible light scattering than as-obtained TiO$_2$. The degradation of crystal violet dye under UVR and simulated solar radiation was significantly slower for TiO$_2$@Y$_2$O$_3$ NPs than TiO$_2$ NPs, with the higher yttria loaded sample (10 wt%) showing the lowest activity. It was suggested that the Y$_2$O$_3$ layer played a role in decreasing the photocatalytic ability of TiO$_2$ by preventing ROS formation by serving as a recombination centre for generated e$^-$/h$^+$ pairs or by impeding their movement, thus inhibiting further reactions with dye molecules on the particle surface. The cytotoxicity of TiO$_2$@Y$_2$O$_3$ composites and TiO$_2$ was studied in vitro by exposing HaCaT cells to the UV filter particles with and without irradiation with simulated solar light. With no simulated solar radiation, cell viability was decreased for TiO$_2$ at concentrations over 100 mg/L. The 5 wt% TiO$_2$@Y$_2$O$_3$ composite decreased cell viability to a lower extent than TiO$_2$, while the 10 wt% TiO$_2$@Y$_2$O$_3$ composite appeared to promote cell viability. For the test with simulated solar radiation, samples were irradiated for either 5 or 10 min and then incubated for 24 h. It was found here that TiO$_2$ NPs and TiO$_2$@Y$_2$O$_3$ composites typically protected cells from damage at the lower radiation dose (5 min) but caused cell death at the higher radiation dose (10 min). Furthermore, both TiO$_2$@Y$_2$O$_3$ composites still showed higher cell viability protection than TiO$_2$ at the different nanoparticle concentrations and radiation doses studied, with higher cell viability observed for the TiO$_2$@Y$_2$O$_3$ composite of higher Y$_2$O$_3$ loading.

Several studies have focused on the modification of TiO$_2$ and ZnO UV filters with cerium oxide (CeO$_2$) NPs due to its UV-attenuation property, as a result of its wide band gap energy of 3.19 eV and radical scavenging ability due to cycling of the Ce^{3+}/Ce^{4+} redox couple within its crystal lattice structure [30–32]. CeO$_2$ is also transparent in the visible region of the spectrum. Mueen et al. [30] obtained ZnO/CeO$_2$ composite NPs (FE-SEM images shown in Fig. 8.1e) through precipitation of CeO$_2$ on commercially available ZnO NPs at pH 9 [29]. Particle sizes of ca. 41, 79, and 90 nm were obtained for composites containing CeO$_2$ at loadings of 10, 5, and 2.5 atomic %, respectively. The UV spectra are presented in Fig. 8.1e, indicating that UV absorption by ZnO, selectively in the UVA region, was enhanced by CeO$_2$ addition and increased with ceria loading. The results of photodegradation studies employing crystal violet dye are presented in Fig. 8.1e, showing that the photoactivity of ZnO/CeO$_2$ composite NPs was substantially lower than that of ZnO NPs and decreased with an increase in CeO$_2$ loading.

Morlando et al. [31] synthesized CeO$_2$-decorated rutile TiO$_2$ nanorods via a precipitation method. CeO$_2$ decoration of TiO$_2$ NPs reduced the photocatalytic activity of rutile TiO$_2$ nanorods by up to 77% and 88% under solar simulated and UV light irradiation, respectively. However, even though the absorption profile of rutile TiO$_2$ nanorods was maintained after CeO$_2$ decoration, the absorption of UVA/B decreased. Morlando et al. [32] later prepared CeO$_2$-decorated commercial TiO$_2$ NPs consisting of a mixture of anatase and rutile crystal phases through precipitation reactions. The CeO$_2$/TiO$_2$ nanocomposites exhibited absorbance of UV primarily in the UVB region but also in the UVA region, though lower than that of pristine TiO$_2$, and visible light transparency in the 400–700 nm region. The extinction coefficients were found to increase with an increase in CeO$_2$ loading from 2.5 to 5% but decreased with a further increase of CeO$_2$ loading to 10%, suggesting that the optimizing of CeO$_2$ loading is important for enhancing the UV attenuation properties of the composite. The nanocomposite containing CeO$_2$ at 5% loading exhibited the highest UV absorption of the composites prepared and near negligible photoactivity towards crystal violet dye degradation under both solar simulated and UV light irradiation. Furthermore, in vitro studies indicated that cell viability of human keratinocyte cells (HaCaT) was not affected by exposure to nanocomposite containing CeO$_2$ at 5% loading over 24 h with and without prior UV exposure.

In contrast, pristine TiO$_2$ NPs exhibited toxicity towards HaCaT cells in certain instances. The decreased photoactivity of CeO$_2$-modified ZnO and TiO$_2$ NPs was attributed to the radical scavenging ability of the CeO$_2$ NPs deposited on the ZnO or TiO$_2$ NPs. The surface defect sites of CeO$_2$ NPs, which typically increase with decreasing particle size, enabled Ce^{3+}/Ce^{4+} cycling, which allowed redox reactions with surface adsorbed species and thus played a role towards hindering ROS generation or quenching any ROS formed.

Chen et al. [33] prepared hollow microcapsules composed of ZnO NPs encapsulated within a CeO$_2$ shell via a sacrificial template method employing polystyrene microspheres, as shown in Fig. 8.1f. For pure ZnO, the UV–vis diffuse reflectance spectra showed a strong absorption band in the UV region extending from 200 to 350 nm. In comparison, the spectra for the ZnO/CeO$_2$ microcapsules showed an absorption band that extended across the entire UV region (200–400 nm) and into the blue-light region (400–480 nm). A thin film of ZnO/CeO$_2$ microcapsules, obtained by spin-coating a dispersion of microcapsules onto a quartz substrate, was of higher optical transparency than a similarly obtained ZnO film, which was attributed to better dispersion of the former in the film. In cytotoxicity studies, the cell viabilities of human skin fibroblast (HSF) cells in contact with ZnO/CeO$_2$ microcapsules remained above 80% even for the highest concentration tested (160 μg/L), while for ZnO NPs, the cell viabilities decreased from 98 to 24% as the concentration increased from 10 to 160 μg/L, attributed to effects of ROS generated by and Zn^{2+} released from ZnO NPs. ZnO/CeO$_2$ microcapsules were found to protect HSF cells from UV-induced damage slightly more than pure ZnO NPs. For this study, a spin coater was used to uniformly spread an alcoholic dispersion of the samples over a quartz slide, which was then mounted over HSF cells in a glass culture dish. Cell viabilities were found to increase with the number of spin-coating layers due to less penetration of UV rays as the layer thickness increased. The authors suggested that the CeO$_2$ shell provided an antioxidative effect by quenching the ROS generated on UV irradiation of ZnO NPs, thus preventing cell damage, as shown in Fig. 8.1f.

Bogusz et al. [34] prepared nanocomposites comprising TiO$_2$ NPs decorated with (BiO)$_2$CO$_3$ clusters of sizes < 10 nm with varying Bi/Ti ratios of 0.02, 0.04 and 0.08. The TiO$_2$/(BiO)$_2$CO$_3$ nanocomposites showed good UVA/B absorption, similar to that of TiO$_2$ NPs, and exhibited lower photocatalytic activity than TiO$_2$ NPs, towards the photodegradation of crystal violet dye solution. The results from in vitro cytotoxicity studies showed the higher biocompatibility of the nanocomposites than TiO$_2$ NPs towards human skin cells (HaCaT) and Madin–Darby dog kidney cells (MDCK). The nanocomposites exhibited cell viabilities above the 71.2% and 61.5%, obtained with TiO$_2$ NPs, for HaCaT and MDCK cells, respectively, at the highest tested concentration tested (500 mg/mL). The cell viabilities of HaCaT cells, under simulated sunlight irradiation, were 64.1 and 74.5% for TiO$_2$ and the nanocomposite (Bi/Ti ratio of 0.08), at 100 μg/mL dose, respectively, suggesting that the nanocomposite showed lower toxicity towards HaCaT cells, possibly due to its ability to reduce photogenerated ROS.

Other studies have focussed on potential inorganic or organic antioxidants to serve as radical scavengers for the ROS generated in sunscreens on UV irradiation of TiO$_2$ and ZnO. Lin et al. [35] investigated fullerenes (C$_{60}$) and nanodiamonds as potential antioxidants in sunscreens containing TiO$_2$. The radical scavenging ability of fullerenes and nanodiamonds increased in the presence of TiO$_2$ under UV irradiation, compared to that of Vitamin C, which showed reduced ability under the same conditions. Prototype sunscreen formulations containing fullerenes (0.1 wt%) or nanodiamonds (0.5 wt%) with TiO$_2$ (8.75 wt%) showed higher photostability over a 4 h UV irradiation period and greater scavenging potential (with and without UVR irradiation), than the TiO$_2$ base cream. The high oxidative stress induced in human dermal fibroblast cells due to the generation of intracellular ROS on irradiation of TiO$_2$ was reduced by nanodiamonds but not by fullerenes. The high SPF value of TiO$_2$ base formulation was not adversely affected by the incorporation of fullerenes (C$_{60}$) and nanodiamonds. The antioxidant activities of C$_{60}$ and nanodiamonds, of a less viscous nature than cream formulations, were maintained for long periods in serums. However, the antioxidant activity of nanodiamonds in the serum decreased significantly after a 3-week shelf-life.

Kanthik et al. [36] prepared sunscreen emulsions with TiO$_2$ particles of three different sizes in combination with Aloe Vera gel extract, a natural product with reported UV attenuation, anti-inflammatory and antioxidant properties. The SPF and UVB absorption edges of the sunscreen emulsions containing only TiO$_2$ increased, while the λ_c and UVA/UVB ratio decreased, with decreasing TiO$_2$ particle size, respectively. The SPF and viscosities of emulsions were increased by an increase in the weight % of TiO$_2$, from 5 to 15 wt%, for a particular TiO$_2$ particle size. The SPF of base emulsions were not affected by the addition of Aloe vera gel extract at varying weight %; however, its addition to TiO$_2$-containing emulsions resulted in slightly increased SPF and decreased viscosity. The best result was obtained for the emulsion containing nanofine TiO$_2$ (at 15 wt%) and Aloe Vera gel extract (at 5%), which had an SPF of ~28, UVA/UVB ratio of ~ 0.63 and λ_c of ~376 nm. However, in this study, the Aloe Vera gel extract did not exhibit UV-attenuation properties. The antioxidant and anti-inflammatory properties imparted by its addition to the TiO$_2$-containing emulsions were not investigated.

The studies presented above have shown that through surface functionalization, coating, or metal ion-doping or decoration with suitable substances, the photocatalytic activity of ZnO and TiO$_2$ can be effectively decreased. The surface functional groups, coatings, and metal ion-dopants or metal-decorations scavenge free radicals or hinder their formation and movement, and in this way, lower the photoactivity of ZnO and TiO$_2$. These modification methods are an achievable strategy for increasing the attractiveness of metal oxides for their continued application as UV filters, which is especially important since ZnO and TiO$_2$ are the only two UV filters currently approved and used in commercial sunscreens. However, relatively high costs and complicated methods for the synthesis of modified ZnO and TiO$_2$, unfavourable for industrial production of UV filters, could hinder their use in sunscreen products.

References

1. Egambaram OP, Pillai SK et al (2020) Materials science challenges in skin UV protection: a review. Photochem Photobiol 96(4):779–797
2. Manaia EB, Kaminski RCK et al (2013) Inorganic UV filters. Braz J Pharm Sci 49(2):201–209
3. Smijs TG, Pavel S et al (2011) Titanium dioxide and zinc oxide NPs in sunscreens: focus on their safety and effectiveness. Nanotechnol Sci Appl 4:95–112
4. Serpone N (2021) Sunscreens and their usefulness: have we made any progress in the last two decades? Photochem Photobiol Sci 20(2):189–244
5. Serpone N, Dondi D et al (2007) Inorganic and organic UV filters: their role and efficacy in sunscreens and suncare products. Inorganica Chim Acta 360(3):794–802
6. Paiva JP, Diniz RR et al (2020) Insights and controversies on sunscreen safety. Crit Rev Toxicol 50(8):707–723
7. Fajzulin I, Zhu X et al (2015) Nanoparticulate inorganic UV absorbers: a review. J Coat Technol Res 12(4):617–632
8. Faure B, Salazar-Alvarez G et al (2013) Dispersion and surface functionalization of oxide nanoparticles for transparent photocatalytic and UV-protecting coatings and sunscreens. Sci Technol Adv Mater 14(2):023001
9. Mitchnick MA, Fairhurst D et al (1999) Microfine zinc oxide (Z-Cote) as a photostable UVA/ UVB sunblock agent. J Am Acad Dermatol 40(1):85–90
10. Carlotti ME, Ugazio E et al (2009) Role of particle coating in controlling skin damage photoinduced by titania nanoparticles. Free Rad Res 43(3):312–322
11. Yin H, Casey PS (2010) Effects of surface chemistry on cytotoxicity, genotoxicity, and the generation of reactive oxygen species induced by ZnO nanoparticles. Langmuir 26(19):15399–15408
12. Bartoszewska M, Adamska E et al (2023) Novelty cosmetic filters based on nanomaterials composed of titanium dioxide nanoparticles. Molecules 28(2):645
13. Choi S, Kim J et al (2022) Plastic-free silica-titania-polyphenol heterojunction hybrids for efficient UV-to-blue light blocking and suppressed photochemical reactivity. Chem Eng J 431(1):133790
14. Choi S, Na H et al (2022) Chitosan-coated mesoporous silica particles as a plastic-free platform for photochemical suppression and stabilization of organic ultraviolet filters. J Photochem Photobiol B 235:112565
15. Morlando A, Sencadas V et al (2018) Suppression of the photocatalytic activity of TiO$_2$ nanoparticles encapsulated by chitosan through a spray–drying method with potential for use in sunblocking applications. Powder Technol 329:252–259
16. Morsella M, D'Alessandro N et al (2016) Improving the sunscreen properties of TiO$_2$ through an understanding of its catalytic properties. ACS Omega 1(3):464–469
17. Kim TH, Park SH et al (2023) A review of biomass-derived UV-shielding materials for bio-composites. Energies (Basel) 16(5):2231
18. Loto AM, Morales JMN et al (2023) Simple preparation of broadband UV filters based on TiO$_2$ coated with aqueous extracts of native trees from the Chaco region of Argentina Photochem Photobiol Sci 22(2):319–331
19. Serpone N, Salinaro A et al (2006) Beneficial effects of photo-inactive titanium dioxide specimens on plasmid DNA, human cells and yeast cells exposed to UVA/UVB simulated sunlight. J Photochem Photobiol A Chem 179(1–2):200–212
20. Abuçafy MP, Manaia EB et al (2016) Gel based sunscreen containing surface modified TiO$_2$ obtained by sol-gel process: Proposal for a transparent UV inorganic filter. J Nanomater 2016:8659240
21. Battistin M, Dissette V et al (2020) A new approach to UV protection by direct surface functionalization of TiO$_2$ with the antioxidant polyphenol dihydroxyphenyl benzimidazole carboxylic acid. Nanomaterials (Basel) 10(2):231

22. Dao DV, van den Bremt M et al (2016) Effect of metal ion doping on the optical properties and the deactivation of photocatalytic activity of ZnO nanopowder for application in sunscreens. Powder Technol 288(2016):366–370
23. Le TH, Bui AT et al (2014) The effect of Fe doping on the suppression of photocatalytic activity of ZnO nanopowder for the application in sunscreens. Powder Technol 268:173–176
24. Ngoc TAN, Ly TQT et al (2023) Sol–gel synthesis of Al-doped TiO$_2$ nanoparticles as UV filters with diminished photocatalytic activity for the application in sunscreen products. J Sol-Gel Sci Technol 108:900–911
25. Khairy M, Zakaria W (2014) Effect of metal-doping of TiO$_2$ nanoparticles on their photocatalytic activities toward removal of organic dyes. Egypt J Pet 23(4):419–426
26. Rampaul A, Parkin IP et al (2007) Damaging and protective properties of inorganic components of sunscreens applied to cultured human skin cells. J Photochem Photobiol A Chem 191(2–3):138–148
27. Reinosa JJ, Docio CMÁ et al (2018) Hierarchical nano ZnO-micro TiO$_2$ composites: high UV protection yield lowering photodegradation in sunscreens. Ceram Int 44(3):2827–2834
28. Parvin N, Mandal TK et al (2023) Application of bimetallic heterojunction nanoparticle-based multishelled porous hollow microspheres as a two-in-one inorganic UV filter. ACS Sustain Chem Eng 11:16133–16143
29. Chaki Borrás M, Sluyter R et al (2020) Y$_2$O$_3$ decorated TiO$_2$ nanoparticles: Enhanced UV attenuation and suppressed photocatalytic activity with promise for cosmetic and sunscreen applications. J Photochem Photobiol B 207:111883
30. Mueen R, Morlando A et al (2020) ZnO/CeO$_2$ nanocomposite with low photocatalytic activity as efficient UV filters. J Mater Sci 55:6834–6847
31. Morlando A, McNamara J et al (2020) Hydrothermal synthesis of rutile TiO$_2$ nanorods and their decoration with CeO$_2$ nanoparticles as low-photocatalytic active ingredients in UV filtering applications. J Mater Sci 55:8095–8108
32. Morlando A, Borrá MC et al (2020) Development of CeO$_2$ nanodot encrusted TiO$_2$ nanoparticles with reduced photocatalytic activity and increased biocompatibility towards a human keratinocyte cell line. J Mater Chem B 8(18):4016–4028
33. Chen F, Ding N et al (2021) Antioxidant hollow structures to reduce the risk of sunscreen. Colloids Surf A: Physicochem Eng 628:127352
34. Bogusz K, Tehei M et al (2018) TiO$_2$/(BiO)$_2$CO$_3$ nanocomposites for ultraviolet filtration with reduced photocatalytic activity. J Mater Chem C 6(21):5639–5650
35. Lin Q, Xu RHJ et al (2019) UV protection and antioxidant activity of nanodiamonds and fullerenes for sunscreen formulations. ACS Appl Nano Mater 2(12):7604–7616
36. Kanthik T, Lokham S et al (2020) Development of sunscreen products containing titanium dioxide and Aloe Vera gel. Key Eng Mater 859:159–165

Chapter 9
Alternative Metal Oxides to ZnO and TiO$_2$ UV Filters

The photocatalytic properties of ZnO and TiO$_2$ have led to renewed concern over their use in sunscreens due to their potentially harmful effects on human health and adverse impacts in the environment linked to this intrinsic property [1, 2]. This has prompted researchers to investigate other metal oxides for UV filters, with similar UV-attenuation properties to TiO$_2$ and ZnO, and their modification to achieve UV filters with improved properties such as lower photocatalytic activity [2–4]. Iron oxide, cerium oxide, and tin oxide are metal oxides investigated to this end [2].

Iron oxides are already common ingredients in cosmetics such as eyeshadows and foundations and in tinted sunscreens due to their brown colour in formulations [5, 6]. Iron oxide is also the most studied filter for visible light (VL) [7]. Though not as harmful as UVR exposure, VL, which comprises close to half of the electro-magnetic radiation in sunlight, has been found to cause various photobiologic effects such as erythema in light-skinned individuals and pigmentation in dark-skinned individuals, and can also induce certain types of photodermatoses including cutaneous porphyrias, solar urticaria, and chronic actinic dermatosis or exacerbate dermatoses such as melasma and post-inflammatory hyperpigmentation [8–10]. A study by Kaye et al. [8] demonstrated that the addition of iron oxide (at 1%) to pastes containing ZnO or TiO$_2$, at 20%, reduced the transmittance of visible light significantly, thus improving the VL photoprotection. Tinted sunscreens that contain inorganic UV filters such as ZnO and TiO$_2$ with iron oxides offer photoprotection against both UVR and VL and are therefore advantageous for individuals afflicted by conditions caused or aggravated by exposure to VL and have enhanced cosmetic acceptability [8–10].

Iron oxides such as hematite (α-Fe$_2$O$_3$), a semiconductor with a band gap energy of 2.2 eV (bulk material), have more recently also been investigated as potential UV filters [2, 6, 11]. Truffault et al. [6] synthesized hematite NPs of the rhombohedral structure via the precipitation method, followed by calcining the product obtained at three different temperatures. The crystallite sizes of the NPs increased with increasing calcination temperature, as confirmed by FE-SEM, TEM and XRD analysis. The

N. H. Kera et al., *Inorganic Ultraviolet Filters in Sunscreen Products*, SpringerBriefs in Materials, https://doi.org/10.1007/978-3-031-64114-5_9

band gap energies of the hematite NPs obtained at calcination temperatures of 300, 400, and 500 °C were calculated to be 3.08, 3.02, and 2.94 eV and thus decreased with temperature. The absorption spectra of the hematite samples contained four bands with maxima at wavelengths of 230, 290, 345, and 395 nm, attributed to Fe^{3+}–O^{2-} charge transfer, with the major peaks occurring between 210 and 230 nm. The three hematite NPs samples absorbed both UVA and UVB radiation though more UVB radiation, with the sample calcined at 500 °C exhibiting the highest UV absorption. The SPF, UVA-PF, and λ_c values of 9.21 ± 1.40, 8.81 ± 1.32, and 390 nm, respectively, determined in vitro for an emulsion containing 10% w/w of hematite sample (calcined at 500 °C), showed that the emulsion met the specifications of a low protection sunscreen, as per the EU classification.

Cardillo et al. [2] found that poly(L-lactic acid) encapsulated Fe$_2$O$_3$ NPs (PLA-Fe$_2$O$_3$) were advantageous over commercially available reference ZnO (Z-Cote HP1, BASF) NPs for application in sunscreen formulations. PLA-Fe$_2$O$_3$ nanocomposite particles exhibited no photocatalytic activity towards the photodegradation of crystal violet dye as opposed to the reference ZnO and TiO$_2$ NPs (Aeroxide P25, Evonik), which exhibited high activity, as can be seen from Fig. 9.1a. The addition of PLA-Fe$_2$O$_3$ (5% w/w) nanocomposite particles to a commercial sunscreen formulation, containing organic filters (25% w/w) and TiO$_2$ (5% w/w), resulted in a SPF value of 50, higher than the SPF of 40 obtained when the reference ZnO NPs (3% w/w) were used instead. The same UVA protection factor was obtained for PLA-Fe$_2$O$_3$ (5% w/w) nanocomposite particles and reference ZnO NPs. The results of cell viability studies, presented in Fig. 9.1a, showed that PLA-Fe$_2$O$_3$ nanocomposite particles were not cytotoxic at concentrations ranging from 6.25 to 50 μg/mL as compared to the reference ZnO NPs which did exhibit cytotoxicity at concentrations of 25 μg/mL and above. In addition, the biodegradable PLA coating also served to reduce the red/brown colour associated with Fe$_2$O$_3$ particles, which arises due to absorption of short wavelength visible light, and which restricts its use to tinted cosmetic and sunscreens products.

Cardillo et al. [11] investigated the effect of cerium doping at varying dopant concentrations on the properties of hematite. In Fig. 9.1b, XRD patterns confirmed that the 5 and 10% cerium-doped hematite samples were primarily composed of the primary phase, hematite. An increase in the dopant concentration above 10% resulted in a decrease in particle size, and the XRD patterns obtained for these samples showed significant peak broadening and a decrease in the intensities of the dominant peaks. This was attributed to the formation of a second phase, CeO$_2$, due to the insolubility of Ce^{4+} within the hematite lattice, starting at dopant levels of 10–20%, which hinders the growth of the primary phase. The morphology of the doped hematite particles also changed from thin, flake-like nanocrystals to spherical NPs as the dopant concentration increased. The absorption spectrum of the 10% cerium-doped sample showed a blue shift and higher absorption of UVR than the pristine hematite and other cerium-doped samples, attributed to interactions involving the effects of the dopant Ce^{4+} ions on the hematite electronic structure and quantum confinement effects due to the decreased particle size (Fig. 9.1b). However, in hematite samples with cerium doping levels above 10%, the formation of secondary phases decreased

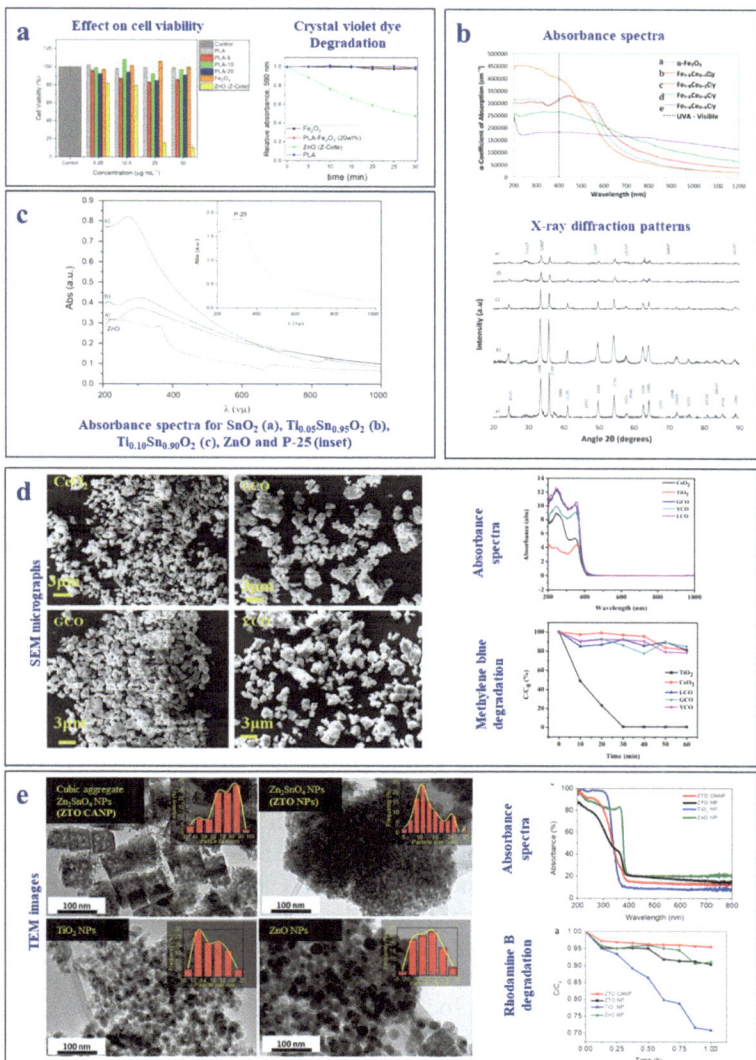

Fig. 9.1 Alternative metal oxides to ZnO and TiO₂ for application as UV filters: **a** Results of dye photodegradation studies and cell viability studies for poly (L-lactic acid) encapsulated Fe₂O₃ NPs [2]. **b** Absorbance spectra and XRD patterns of cerium-doped hematites [11]. **c** Absorbance spectra of titanium doped tin dioxide nanopowders [3]. **a, b,** and **c** Reproduced with permission from Elsevier Science Ltd. **d** SEM images and absorbance spectra of rare-earth metal-doped cerates and results of dye photodegradation studies. Reproduced with permission from [19]. Copyright 2022, Wiley. **e** TEM images and absorbance spectra of Zn₂SnO₄ NPs and results of dye photodegradation studies. Reproduced with permission from [20]. Copyright 2011, National Institute for Materials Science in partnership with Taylor & Francis Group

these advantageous effects. The band gap energy of the 10% cerium-doped sample (2.51 eV) was also higher than that of the pristine hematite (2.42).

Morlando et al. [3] synthesized titanium-doped tin dioxide nanopowders in a simple co-precipitation method. Successful doping of tin dioxide by titanium and the tetragonal rutile phase of tin dioxide obtained was confirmed by XRD analysis, which, with TEM analysis, also showed that particle size decreased with increasing dopant levels. The titanium-doped tin dioxide nanopowders, with higher band gap energies, displayed significantly lower photocatalytic activity towards the degradation of methylene blue (MB) dye than commercial ZnO and TiO_2 NPs (P-25). The UV–visible spectra (Fig. 9.1c) showed that the absorption of UVR by the titanium-doped tin dioxide nanopowders was higher than that of the commercial ZnO but lower than that of the uncoated commercial TiO_2 NPs.

Cerium oxide (CeO_2) has garnered interest as an alternative to ZnO and TiO_2 UV filters in sunscreens due to its UV absorption properties (band gap energy of 3.2 eV), high visible light transparency, low toxicity and relatively low photocatalytic activity [4, 12, 13].

Zholobak et al. [14] showed that the UV-attenuation properties of citrate-stabilized ceria colloid solutions, in terms of SPF values, critical absorption wavelengths, and UVA/UVB-ratio, were similar to that of solutions of TiO_2 and ZnO NPs. The λ_c of the ceria solution, TiO_2 solution, and ZnO solutions, each at 1% concentrations, were 355 nm, 364 nm, and 379 nm, respectively. The UVA/UVB ratio obtained for the ceria solution (1%) of 0.28 was close to that of the TiO_2 solution (0.36 nm) but much lower than that of the ZnO solution (0.86 nm) at the same concentration. A sunscreen formulation of SPF 30 could be obtained with 19% TiO_2, 27% CeO_2 or 42% ZnO NPs. The photocatalytic activity of ceria colloid solution and ceria NPs towards the photodegradation of methyl orange was substantially lower than that of TiO_2. Colloid ceria solutions were not cytotoxic to mouse fibroblast cells (L929) and mammalian fibroblast-like cells (VERO) in the absence of UVR and protected cells from damage caused by exposure to UVR. The authors suggested that ceria NPs afforded cells protection from indirect damage by inactivating the ROS, nitric oxide and peroxide lipid oxidation products generated on UVR irradiation. In cells treated with ceria NPs prior to irradiation, protection was due to the increase of superoxide dismutase 2 (SOD2) production, an enzyme responsible for converting highly reactive superoxide to the much less reactive hydrogen peroxide (H_2O_2), brought about by the ceria NPs.

Caputo et al. [13] prepared TiO_2 (anatase structure) and CeO_2 NPs (fluorite structure) for a comparative study by using a wet chemistry method that involved heating their respective precursors at 450 °C. Both TiO_2 and CeO_2 NPs are absorbed in the UV region, with TiO_2 showing a higher absorption capability than CeO_2 NPs. As such, TiO_2 NPs protected human keratinocyte cells (HaCat) from damage caused by exposure to UVB radiation to a greater extent than CeO_2 NPs. However, TiO_2 exhibited higher toxicity and mutagenic effects than CeO_2 after UV irradiation, with more pronounced effects observed for UVA irradiation due to its penchant to generate highly toxic hydroxyl radicals on being irradiated. CeO_2 NPs, in the absence of UV radiation, did not exhibit toxicity or increase the toxic effects caused by UVR.

This study also found that the antioxidant effects of CeO$_2$ NPs, arising from the Ce$^{3+/4+}$ redox switch, contributed to preventing mutagenesis caused by UVR by decreasing DNA damage and improving the rate of DNA repair. Furthermore, CeO$_2$ NPs were more efficient than molecular antioxidants (Trolox and NAC) used in commercial sunscreens at scavenging the ROS generated on irradiation of TiO$_2$ NPs. It was suggested that CeO$_2$ NPs could be incorporated into traditional TiO$_2$-based sunscreen formulations to add antioxidant properties and reduce undesirable effects arising due to the photocatalytic nature of TiO$_2$.

Boutard et al. [15] prepared spherical CeO$_2$ NPs of pure fluorite structure using a microwave-hydrothermal method. The in vitro determined SPF value of 49 of an emulsion containing a mixture of TiO$_2$ and organic sunscreens and CeO$_2$ (3 wt%) was higher than the SPF value of 43 achieved with the same base emulsion containing ZnO (3 wt%). The cytotoxicity of the prepared CeO$_2$ NPs was compared to that of five ZnO-based sunscreens in the market using MTT assays. The cell viability of NCTC2544 human keratinocytes cells was not affected by CeO$_2$ NPs but was decreased by exposure to ZnO in a time and dose-dependent manner.

However, as for ZnO and TiO$_2$, the use of CeO$_2$ for photoprotection applications is limited by its inherent photocatalytic activity. The fluorite structure of CeO$_2$ is unstable as the ionic radius of Ce^{4+} is not sufficiently large to attain the ideal ionic radius ratio for a metallic element in an oxide of MO8 coordination. This results in the tendency of Ce^{4+} to easily transform into Ce^{3+}, with a larger ionic radius, a reaction in which oxygen is released which contributes to the photocatalytic activity of CeO$_2$.

Several studies have focused on the doping of CeO$_2$ to reduce its photocatalytic activity and band gap energy values [12]. The fluorite structure of CeO$_2$ can be stabilized by replacing Ce^{4+} with cations of larger ionic radius and lower valence state.

Yabe et al. [16] found that the doping of ceria with Ca^{2+} (20 mol%) and Zn^{2+} (20 mol%) reduced the catalytic activity of CeO$_2$ NPs towards the oxidation of castor oil in air, and the doped ceria NPs also showed good absorption of UVR when compared to undoped ceria. Doped and undoped ceria showed substantially lower photocatalytic activity towards the decomposition of phenol than TiO$_2$, possibly due to the presence of oxygen defects in their structures. The Ca^{2+} (20 mol%)-doped ceria NPs were also highly transparent to visible light and, therefore, advantageous for use as UV filters in sunscreens. However, the small particle size of Ca^{2+} (20 mol%)-doped ceria NPs, of 2–4 nm, raises concerns over their use in sunscreens.

Truffault et al. [12] modified the optical and structural properties of CeO$_2$ NPs by doping with Ca, with dopant concentrations varying between 10 and 50 mol%. The Ca-doped CeO$_2$ samples with dopant concentrations below 20% were composed of one phase. Of the different doped samples, the Ca-doped CeO$_2$ (10 mol%) was found to be most efficient for UV absorption, absorbing UVR maximally between 265 and 325 nm. The samples prepared with dopant concentrations of 30% and above contained a CaCO$_3$ phase in addition to CeO$_2$, rendering these unsuitable for use in cosmetics. Ca doping of CeO$_2$ caused a blue shift of the absorption spectrum of undoped CeO$_2$, advantageous for filtering short UVA. Truffault et al. [17] further investigated the potential of Ca-doped CeO$_2$ to replace ZnO in a classical sunscreen formulation used commercially. In this study, 10% Ca-doped CeO$_2$ was synthesized by two methods: the co-precipitation method and spray pyrolysis. The photoprotection efficiency of three water-in-oil emulsions containing commercial TiO$_2$ (19%) with either commercial ZnO (classical formulation) or 10% Ca-doped CeO$_2$ synthesized by spray pyrolysis or 10% Ca-doped CeO$_2$ synthesized by co-precipitation, each at 6% by mass, were assessed in vitro. With SPF values above 30 and critical wavelengths higher than 370 nm, all three emulsions could be classified as broad-spectrum, high-protection sunscreens. The two emulsions containing 10% Ca-doped CeO$_2$ offered better UVB and UVA protection than the classical emulsion. The SPF of the sample synthesized by co-precipitation increased by 27%, while the UVA-PF of the sample synthesized by spray pyrolysis increased by 19% when compared to the classical emulsion. The photoprotection efficiency of the Ca-doped CeO$_2$ samples was dependent on the size and shape of the particles and, therefore, affected by the synthesis method employed.

Truffault et al. [18] later synthesized iron-doped CeO$_2$ NPs, with dopant concentrations from 10 to 30 mol%, using a co-precipitation reaction and calcining the product obtained thereof at 500 °C for 6 h. Iron was selected as it has a lower valence state than Ce^{4+} and because of its biocompatibility. XRD analysis indicated that the 10 mol% iron-doped samples contained one phase, pure CeO$_2$, while the iron-doped CeO$_2$ NPs with iron dopant concentrations of 20 mol% or more also contained hematite, suggesting that the solubility limit for Fe in the CeO$_2$ lattice was between 10 and 20 mol% doping levels. The mean crystallite size decreased with the iron doping level. The absorption curves of pure CeO$_2$ and 10 mol% iron-doped CeO$_2$ both showed maxima in the UVB region. However, the 10 mol% doped CeO$_2$ curve showed a blue shift of the absorption maximum, associated with the particle size decrease due to doping. However, the colour of the iron-doped CeO$_2$, ranging from light to dark brown depending on the iron concentration, may limit their use to tinted sunscreens.

Raj et al. [19] prepared rare-earth metal-doped cerates with the general formula RE$_2$Ce$_2$O$_7$, where RE = La, Gd, and Y, through a conventional ceramic route. The SEM micrographs, as presented in Fig. 9.1d, of the La$_2$Ce$_2$O$_7$ (LCO), Gd$_2$Ce$_2$O$_7$ (GCO) and YCO for Y$_2$Ce$_2$O$_7$ (YCO) samples showed particles with uniform sizes ranging between 1–2 μm. XRD analysis confirmed the cubic fluorite structure of the rare-earth metal-doped cerates and lattice expansion on doping. The UV spectra obtained, as shown in Fig. 9.1b, for the doped cerates showed intense UV absorption in the UVA/B region, higher than that of TiO$_2$ and CeO$_2$, attributed to an increase in the polarizability of oxygen anions bonded to cerium, the order La > Gd > Y ions, as a result of doping. The polarizability of oxygen anions depended on the ionic size and electronegativity of the dopant ions. Furthermore, the photoactivity of the doped cerates towards the degradation of methylene blue dye under UV-irradiation, as shown in Fig. 9.1d, was substantially lower than that of TiO$_2$ but close to that of CeO$_2$, as a result of the presence of oxygen defects, at similar concentrations, in both doped and undoped CeO$_2$.

Al-Attafi et al. [20] investigated a ternary metal oxide, zinc stannate (Zn_2SnO_4), as a potential alternative UV filter and compared its properties to that of ZnO and TiO_2. Two different zinc stannate morphologies were synthesized via solvothermal methods followed by calcination steps, including zinc stannate nanoparticles of particle sizes 8 ± 2 nm and cubic aggregate zinc stannate nanoparticles, consisting of aggregates of particles of sizes 75 ± 15 nm, as presented in Fig. 9.1e. ZnO NPs and TiO_2 NPs with particle sizes of 22 ± 5 nm and 16 ± 4 nm, respectively, were also prepared for comparison purposes. The UV–visible spectra (Fig. 9.1e) obtained showed that the UVA absorption range for zinc stannate NPs was broader than that of TiO_2 NP but narrower than that of ZnO NPs; however, the absorption coefficients were lower for zinc stannate NPs than ZnO and TiO_2 NPs in the UV region. The zinc stannate NPs showed higher reflectance in the UV region than ZnO and TiO_2 NPs. In the visible region of the spectra, the zinc stannate NPs showed reflectances similar to that of ZnO but higher than TiO_2, which exhibited low reflectance. The bandgap energy values obtained from Tauc plots for zinc stannate NPs (3.25 eV) were similar to those obtained for ZnO NPs (3.2 eV) and TiO_2 NPs (3.3 eV). In Rhodamine B dye degradation studies, the zinc stannate NPs exhibited photocatalytic activity lower than TiO_2 NPs but similar to that of ZnO NPs, as shown in Fig. 9.1e. High cell viabilities were maintained (above 80%) for MDAMB231 cells in contact with zinc stannate and TiO_2 NPs in the dark at all concentrations studies (0–200 µg/mL) which indicate low inherent toxicities. This was in contrast to ZnO NPs, for which cell viabilities decreased sharply (~50% for NP concentration of 25 µg/mL) and were close to zero for NP concentrations above 50 µg/mL. However, cell viability studies were not carried out under UV irradiation conditions, the results of which may have indicated effects on cells and the levels of ROS generated.

Studies in the literature have indicated the potential of iron and cerium oxides for use as UV filters in sunscreens as alternatives to ZnO and TiO_2. Through modification by coating and metal ion doping, the properties of these oxides were improved for photoprotection applications, as for ZnO and TiO_2. However, the use of iron and cerium oxides in sunscreen products is subject to their approval for use as UV filters by regulatory bodies, which is unlikely in the near future, and their higher relative cost, especially in the case of cerium oxide, may also hinder their use.

References

1. Scientific Committee on Consumer Safety, Chaudhry Q (2015) Opinion of the Scientific Committee on Consumer Safety (SCCS)—revision of the opinion on the safety of the use of titanium dioxide, nano form, in cosmetic products. Regul Toxicol Pharmacol 73(2):669–670
2. Cardillo D, Sencadas V et al (2021) Attenuation of UV absorption by poly(lactic acid)-iron oxide nanocomposite particles and their potential application in sunscreens. Chem Eng J 405:126843
3. Morlando A, Cardillo D et al (2016) Titanium doped tin dioxide as potential UV filter with low photocatalytic activity for sunscreen products. Mater Lett 171:289–292

4. Parwaiz S, Khan MM et al (2019) CeO$_2$-based nanocomposites: an advanced alternative to TiO$_2$ and ZnO in sunscreens. Mater Express 9(3):185–202
5. Serpone N, Dondi D et al (2007) Inorganic and organic UV filters: their role and efficacy in sunscreens and suncare products. Inorganica Chim Acta 360(3):794–802
6. Truffault L, Choquenet B et al (2011) Synthesis of nano-hematite for possible use in sunscreens. J Nanosci Nanotechnol 11(3):2413–2420
7. Ma Y, Yoo J et al (2021) History of sunscreen: an updated view. J Cosmet Dermatol 20(4):1044–1049
8. Kaye ET, Levin JA et al (1991) Efficiency of opaque photoprotective agents in the visible light range. Arch Dermatol 127(3):351–355
9. Geisler AN, Austin E et al (2021) Visible light. Part II: photoprotection against visible and ultraviolet light. J Am Acad Dermatol 84(5):1233–1244
10. Lyons AB, Trullas C et al (2021) Photoprotection beyond ultraviolet radiation: a review of tinted sunscreens. J Am Acad Dermatol 84(5):1393–1397
11. Cardillo D, Konstantinov K et al (2013) The effects of cerium doping on the size, morphology, and optical properties of α-hematite nanoparticles for ultraviolet filtration. Mater Res Bull 48(11):4521–4525
12. Truffault L, Ta MT et al (2010) Application of nanostructured Ca doped CeO$_2$ for ultraviolet filtration. Mater Res Bull 45(5):527–535
13. Caputo F, De Nicola M et al (2015) Cerium oxide nanoparticles, combining antioxidant and UV shielding properties, prevent UV-induced cell damage and mutagenesis. Nanoscale 7(38):15643–15656
14. Zholobak NM, Ivanov VK et al (2015) UV-shielding property, photocatalytic activity and photocytotoxicity of ceria colloid solutions. J Photochem Photobiol B 102(1):32–38
15. Boutard T, Rousseau B et al (2013) Comparison of photoprotection efficiency and antiproliferative activity of ZnO commercial sunscreens and CeO$_2$. Mater Lett 108:13–16
16. Yabe S, Sato T (2003) Cerium oxide for sunscreen cosmetics. J Solid State Chem 171(1–2):7–11
17. Truffault L, Winton B et al (2012) Cerium oxide based particles as possible alternative to ZnO in sunscreens: effect of the synthesis method on the photoprotection results. Mater Lett 68:357–360
18. Truffault L, Yao QW (2011) Synthesis and characterization of Fe doped CeO$_2$ nanoparticles for pigmented ultraviolet filter applications. J Nanosci Nanotechnol 11(5):4019–4028
19. Raj AKV, Rao PP (2022) Intense UV absorbers in fluorite-type rare-earth cerates for sunscreen formulations. J Am Ceram Soc 105(12):7366–7373
20. Al-Attafi K, Al-Keisy et al (2023) Zn$_2$SnO$_4$ ternary metal oxide for ultraviolet radiation filter application: a comparative study with TiO$_2$ and ZnO. Sci Technol Adv Mater 24(1):2277678

Chapter 10
Alternative Inorganic UV Filters

Certain features of TiO$_2$, ZnO, and other semiconducting metal oxides, especially their NP forms, such as small particle sizes and high photoreactivity, have renewed concerns over their use in sunscreens due to their potential risks to human health [1, 2]. Recently, the SCCS has dissuaded the use of the highly photocatalytic form of TiO$_2$, anatase, and further recommended that its concentration should not exceed 5% of the total TiO$_2$ in a sunscreen formulation due to safety concerns [2, 3]. Even though ZnO and less photocatalytic forms of TiO$_2$, such as rutile, are still recommended for use in sunscreens at concentrations up to 25%, this could change as more information comes to light regarding the harmful effects of semiconductor-based UV filters and/ or NPs in the human body or to organisms in the environment. This has resulted in the search for alternative inorganic substances for use as UV filters with good UVR attenuation and without the undesirable features of metal oxides [1, 2]. Inorganic substances investigated as alternatives to TiO$_2$, ZnO, and other metal oxides for use as UV filters include talc, kaolin, red veterinary petrolatum, ichthammol, talc, calamine, carbonate and phosphate-based nanomaterials, and pristine and modified hydroxyapatites and hydrotalcites, among others [1, 2, 4]. Natural materials are of particular interest for use as UV filters, due to their relatively low cost, their abundance in nature, nontoxicity and high biocompatibility which lends to a low risk of harming human health or causing adverse environmental effects [5].

Hydroxyapatite (Ca$_{10}$(PO$_4$)$_6$(OH)$_2$) (HAp) is a mineral that occurs naturally as a component of human and animal bones and has been studied as an alternative to conventional inorganic UV filters due to its nontoxicity, high biocompatibility, and low cost [2, 6].

Amin et al. [7] synthesized HAp-ascorbic acid nanocomposites for potential use in sunscreens in a method employing PVP for dispersion and stabilization. TEM imaging showed that rod-shaped particles were obtained, the morphology likely facilitated by the PVP present, with thicknesses ranging from 20–30 nm and lengths above 150 nm. The absorption spectrum of HAp-ascorbic acid nanocomposites contained absorption peaks in the UVC region at 225 and 250 nm. The cell viabilities of human

N. H. Kera et al., *Inorganic Ultraviolet Filters in Sunscreen Products*,
SpringerBriefs in Materials, https://doi.org/10.1007/978-3-031-64114-5_10

dermal fibroblast cells (SKIN) and human epidermal keratinocyte cells (HaCaT) subjected to UV irradiation ($\lambda_{max} = 254$ nm) were stimulated when treated with HAp-ascorbic acid nanocomposites, relative to the control. ROS generation in cells on UV irradiation was significantly decreased by treatment with HAp-ascorbic acid nanocomposites, attributed to the scavenging of ROS by the ascorbic acid present.

Morsy et al. [8] synthesized HAp-chitosan gel for application as an antibacterial sunscreen by using a coprecipitation method. SEM imaging showed that the composite was comprised of roughly spherical particles (of 200 nm diameter) of HAp encapsulated by a chitosan matrix. The HAp-chitosan gel exhibited antibacterial activity against different multidrug-resistant skin bacteria and absorbed UV radiation of wavelengths below 350 nm, with increasing absorbance intensity as the wavelength decreased towards 200 nm.

The modification of hydroxyapatite by doping with different cations such as Fe^{2+}, Mn^{2+}, Zn^{2+}, Cr^{3+}, and Fe^{3+} has been investigated to improve its UV-absorbing properties [2, 6]. Piccirillo et al. [5, 9] prepared iron-doped HAp by calcination of waste cod fish bones treated with an iron (II) chloride solution. XRD analysis showed that the product contained mainly calcium hydrogen iron phosphate ($Ca_9FeH(PO_4)_7$), HAp, and a small amount of hematite. The UV–visible spectra (Fig. 10.1a) showed that iron-doped HAp samples had broader UV absorption ranges than commercial ZnO and TiO_2. However, the iron-doped HAp samples also absorbed strongly in the visible region, unlike ZnO and TiO_2. The radicals generated on UV irradiation of samples were monitored by reacting these species with ABTS and measuring absorbances at $\lambda_{max} = 734$ nm. The results (Fig. 10.1a) showed that iron-doped HAp samples did not generate radicals on UV irradiation, in contrast to ZnO and TiO_2. The sunscreen cream formulations containing the iron-doped HAp, at 1–20% concentrations, were categorized as broad-spectrum based on the obtained λ_c values, typically above 370 nm, and high UVA/UVB ratios (typically above 0.9). The sunscreen cream formulation containing 15% iron-doped HAp did not cause any irritation or erythema when applied to the skin (Fig. 10.1a). Even though the SPF values of two sunscreen cream formulations containing 15 wt % iron-doped HAp with varying wt% of iron were not high, 3.49 and 4.08, the SPF of an emulsion containing TiO_2 (7.5 wt%) was significantly increased from 4.98 to 8.29 by incorporating iron-doped HAp at 7.5 wt%, suggesting a synergistic effect [8]. However, the emulsions containing iron-doped HAp with or without TiO_2 had colours ranging from medium to dark reddish-brown, as shown in Fig. 10.1a, which may limit their uses to tinted sunscreens or other cosmetics. Similar findings were obtained in a study by Hadagalli et al. [10] that focussed on Fe^{3+} ionic substitution of Ca^{2+} in hydroxyapatite, obtained from cuttlefish bone, for imparting UV absorption properties to HAp, which does not absorb in the UV region of the spectrum in its pristine state. UV–visible spectra (Fig. 10.1b) showed that Fe-doped HAp (Fe-HA) powders absorbed across the UV range and in the visible region, unlike undoped HA powder, which did not absorb at all in the UV and visible regions. Fe^{3+} doping of HAp, at concentrations varying from 0.01 M to 0.05, resulted in a decrease in the unit cell volume of HAp, indicating lattice contraction, due to the smaller size of Fe^{3+} than Ca^{2+}, and decreased the band gap energy value of HAp from 5.9 eV to within the 1.84–2.18 eV range, obtained for the different Fe-HA powders.

Araujo et al. [11] synthesized pure, Zn-doped, and Mn-doped HAp and trical-cium phosphate (β-TCP) via a wet chemical process involving the precipitation of product on the addition of a calcium nitrate solution (also containing zinc nitrate or manganese nitrate for doped samples) to a diammonium phosphate solution, at appropriate stoichiometric levels and solution pH. The precipitates were calcined at 500 °C for 1 h to obtain pure and doped HAp and at 800 °C for 2 h to obtain pure and doped β-TCP. XRD analysis confirmed the Zn and Mn doping of HAp and β-TCP and the presence of secondary phases such as CaO in all samples, ZnO or MnO in doped samples and HAp in doped β-TCP samples. Of the three β-TCP samples, only Mn-doped β-TCP displayed UV absorption, with maxima in the 200–245 nm and 252–386 regions (Fig. 10.1c). The UV absorption range of HAp, 200–340 nm, changed on doping with Zn to 213–400 nm, and on doping with Mn, to 200–388 nm, resulting in enhanced UVA absorption (Fig. 10.1c). However, Mn-doped HAp showed increased absorption in the visible region that could impart an undesirable colour to sunscreen formulations.

Pyo et al. [12] synthesized silver-doped HAp via a cation exchange reaction, and annealing of the product recovered for 3 h at 100 °C [11]. The HAp was prepared in the laboratory using a hydrothermal method with Na_2HPO_4 and $Ca(NO_3)_2.4H_2O$ as the starting materials. XRD patterns obtained for the sample contained peaks attributed to HAp and metallic silver, while TEM images (Fig. 10.1d) showed silver NPs present on the rod-shaped HAp particles. The silver-doping of HAp enhanced its absorption in both the UV and visible light regions of the spectrum, in Fig. 10.1d, with maximum absorption occurring at a wavelength of 509 nm.

Abbas et al. [13] investigated calcium silicate, a mineral occurring in soil, as an alternative UV filter for use in sunscreens due to its lower refractive index and photocatalytic activity than ZnO and TiO_2. Sunscreen formulations based on oil/water emulsions were prepared containing an organic UV filter, octyl methoxycinnamate (OMC) (at 5 wt%), and different inorganic UV filters (at 5 wt%), for comparison purposes, as presented in Fig. 10.1e. The formulation containing calcium silicate had an SPF value, as calculated from the Mansur equation, of 38 compared to that of formulations containing no inorganic UV filter, ZnO and TiO_2, with SPF values of ~12, ~25 and ~48, respectively. The UV spectrum of the sunscreen formulation with calcium silicate (Fig. 10.1e) contained a sharp, narrow peak in the 260–340 region with λ_{max} at 290 nm. No adverse effects were observed on UV irradiation of the skin of hairless mice after the application of the sunscreen formulation containing calcium silicate, and calcium absorption in the skin was also negligible.

Layered double hydroxides (LDHs) are ionic solids that have a two-dimensional layered structure consisting of layers of divalent/trivalent metal cations coordinated to hydroxide anions (OH^-) [14]. The metal cations impart a positive charge to the layers in the structure, which is balanced by the presence of intercalated water molecules or anions in the interlayers between the layers. Egambaram et al. [14] prepared a Zn-Ti-layered double hydroxide (Zn-Ti LDHs) via a hydrothermal synthesis method with the aim of obtaining a UV filter of greater structural and photostability than

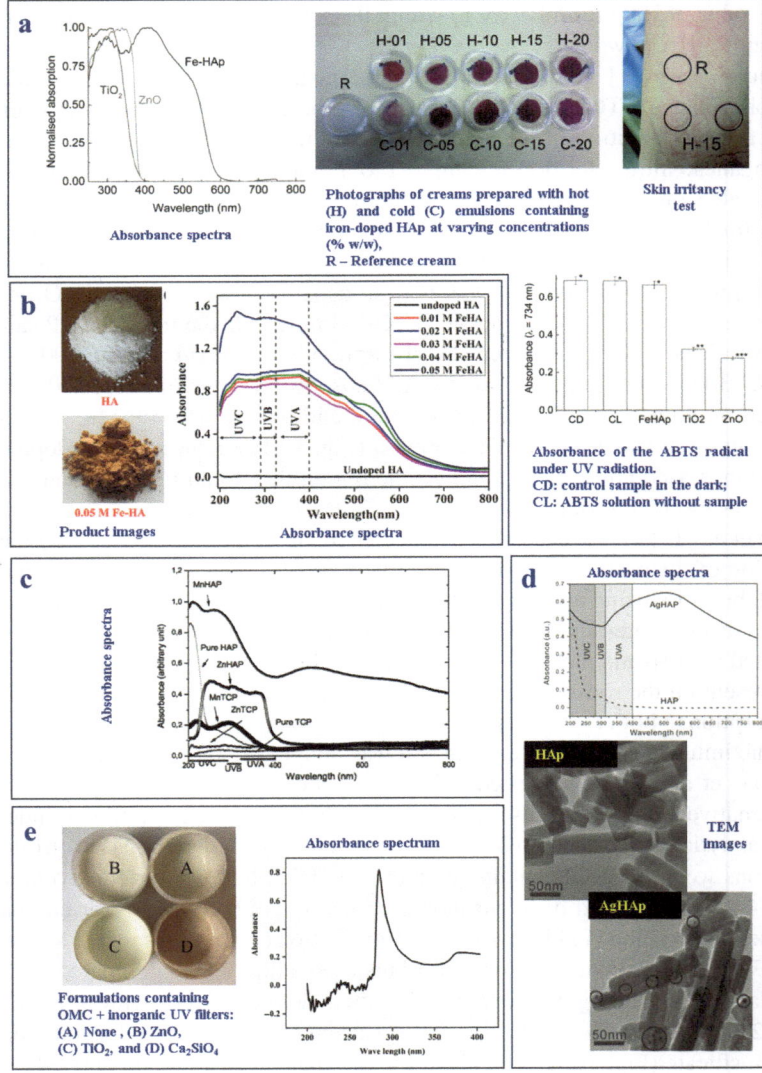

Fig. 10.1 Alternative inorganic materials investigated for application as UV filters: **a** UV absorption spectrum of Fe-doped HAp-based material, photograph of the emulsions prepared containing iron-doped HAp, and results from the skin irritancy and photoactivity tests. Reproduced with permission from [5]. Copyright 2014, Royal Society of Chemistry. **b** UV absorption spectra of Fe-doped HAps with varying Fe content and material photographs [10]. **c** UV absorbance spectra of Zn- and Mn-doped phosphates. Reproduced with permission from [11]. Copyright 2014, Elsevier. **d** UV absorbance spectra and TEM images of Ag-doped HAp [12]. **e** UV absorbance spectra of calcium silicate and photographs of formulations [13]. **b, d, e** Reproduced with permission from Wiley

conventional ZnO and TiO$_2$. The diffuse reflectance spectra showed that Zn-Ti LDH showed a better reflection of UVR than TiO$_2$ and ZnO but exhibited lower UVR absorption. Zn-Ti LDH displayed lower photoreactivity towards the degradation of methylene blue dye when subjected to both UVA (365 nm) and UVB (254 nm) irradiation than TiO$_2$ and ZnO. These results were correlated to the higher band gap energy value of Zn-Ti LDH of 3.72 eV than that of TiO$_2$ (3.23 eV) and ZnO (3.27 eV), as obtained from Tauc plots and the structural characteristics of Zn-Ti LDH. An SPF value of 17.9 obtained for Zn-Ti LDH, as calculated from the UV transmittance spectra using Mansur's Equation, showed that the material was suitable for UVB attenuation.

Egambaram et al. [15] further investigated LDHs for sunscreen applications and characterized Zn-Al and Mg-Al LDHs in addition to Zn-Ti LDH. The incorporation of different metal cations into the LDH structure affected their optical and other properties. The diffuse reflectance spectra of the three LDHs studied showed that Zn-Ti LDH showed better reflectance of both UVA and UVB radiation than Zn-Al and Mg-Al LDHs. This was attributed to a synergistic effect arising due to the presence Zn^{2+} and Ti^{4+} in the structure of the LDH. The values of the band gap energies of the studied materials, in increasing order, were TiO$_2$ < ZnO < Zn-Ti LDH < Zn-Al LDH < Mg-Al LDH. The observed photocatalytic activity of the materials studied, in terms of methylene blue degradation, appeared to be related to their band gap energy values. As such, the photoreactivity of Zn-Ti LDH was higher than Zn-Al LDH and Mg-Al LDH but lower than ZnO and TiO$_2$. The in vitro determined SPF, UVA-PF, and λ_c values of formulations containing Zn-Ti LDH and Zn-Al LDH, at 2% levels, were 6.11 and 4.29, 2.65 and 3.82, and 384 nm and 387 nm, respectively. Based on these values, the formulations containing Zn-Ti LDH and Zn-Al LDH met the criteria to be classified as broad-spectrum sunscreens providing UVA and UVB protection.

LDHs were also investigated as vehicles for the encapsulation of organic UV filters [16]. Most organic UV filters are prone to photodegradation, which not only reduces their UV attenuation properties but also generates toxic byproducts that can cause contact dermatitis, irritation, and other adverse effects on the skin [16–19]. The intercalation of organic UV filters in LDHs is to improve their photostability by hindering their release into the sunscreen formulations, thus preventing the adverse effects associated with their direct interaction with and penetration into the skin while still maintaining their UV attenuation properties [16–19]. Organic UV filters with anionic forms such as p-aminobenzoic acid (PABA) and benzophenone and its derivatives are suitable for LDH intercalation [16–19]. The instability of PABA in UV light and its highly photosensitizing nature, which has led to its disuse in commercial sunscreens, and the potential toxicity and carcinogenicity of benzophenone derivatives make these organic UV filters ideal candidates for LDH intercalation [16, 17, 19]. Perioli et al. [17] prepared PABA-intercalated MgAl– and ZnAl–LDHs via an anion exchange method. ZnAl–LDHs–PABA absorbed more UVA/B radiation than free PABA, while MgAl–LDHs–PABA showed lower absorption than both. In addition, ZnAl–LDHs protected the intercalated PABA from photodegradation, thereby increasing its photostability. Perioli et al. [18] made similar observations in a later study where the photostability of 5-benzoyl-4-hydroxy-2-methoxy-benzenesulphonate acid (4BHF), an organic UV filter, was improved by its intercalation into ZnAl-LDHs.

Furthermore, the spectral features and UV attenuation properties of the 4BHF-intercalated ZnAl-LDHs were retained after incorporation into silicone cream-based sunscreen formulations. The absorption spectra of the 4BHF-intercalated ZnAl-LDHs differed slightly depending on the form of 4BHF intercalated, either mono-anionic or di-anionic, due to the bi-protic nature of 4BHF. In both studies, the release of organic UV filter (PABA or 4BHF) from silicone cream-based sunscreen formulations into water, phosphate buffer, and simulated seawater mediums was significantly lower from the formulations containing organic filter-intercalated LDHs than from those containing the unprotected organic molecules. However, the release of the organic UV filters from the intercalated hydrotalcites was slightly higher in phosphate buffer (pH 5.5) than in water or simulated seawater. This suggested that the phosphate anions penetrated the cream formulations, entered the interlayer spaces, and caused the release of 4BHF via ion exchange mechanisms. Further stability studies are therefore necessary to determine the release of organic UV filters from intercalated hydrotalcites within sunscreen formulations over time.

Similarly, Abdul Aziz et al. [19] synthesized PABA-intercalated zinc LDHs using a co-precipitation method with commercial ZnO as a precursor. The structural, thermal, and surface properties of the material were characterized using different techniques such as powder X-ray diffraction, FTIR spectroscopy, and TGA analysis. However, the optical properties of the materials and their performance as UV filters were not evaluated.

Different zeolites have also been investigated for the encapsulation of organic UV filters, including avobenzone and octinoxate, in a number of studies [20, 21]. Enhanced UV attenuation was observed for hybrids of organic UV filters encapsulated by cationic, unprotonated zeolites, while hybrids incorporating high silica zeolites exhibited low UV attenuation. Hybrids incorporating protonated zeolites exhibited UV attenuation properties in between that of hybrids containing cationic, unprotonated zeolites and those containing high silica zeolites. Encapsulation in zeolites improved the photostability of organic UV filters due to guest–host stabilization, which potentially prevents contact between the skin and UV filters.

In a study by Lademann et al. [22] silica microparticles of 500 nm particle size were found to increase the SPF of an organic UV filter-based sunscreen formulation by 1.4 times. This was an interesting finding since silica microparticles do not inherently absorb UVA and/or UVB radiation. The observed increase in SPF was attributed to the visible light scattering properties of silica microparticles. The scattering of light photons by silica microparticles increases the optical pathway of photons in the formulation medium, which results in increased photon absorption by the organic UV filters.

Different inorganic materials have been investigated as alternatives to metal oxide-based UV filters. Natural materials are advantageous for use in sunscreens due to their abundance, biocompatibility and low photoreactivity, but they require modification to improve their UV attenuation properties. Therefore, their use in sunscreens may be hindered by their lower efficiency, even after modification, when compared to traditional inorganic UV filters, due to the high SPFs demanded by consumers and regulatory bodies, complex procedures for their modification and other undesirable features such as their coloured nature and nontransparency in the visible region. Their use in commercial products in the near future will also depend on their regulation.

References

1. Manaia EB, Kaminski RCK et al (2013) Inorganic UV filters. Braz J Pharm Sci 49(2):201–209
2. Nery ÉM, Martinez RM et al (2021) A short review of alternative ingredients and technologies of inorganic UV filters. J Cosmet Dermatol 20(4):1061–1065
3. Scientific Committee on Consumer Safety, Chaudhry Q (2015) Opinion of the Scientific Committee on Consumer Safety (SCCS)—revision of the opinion on the safety of the use of titanium dioxide, nano form, in cosmetic products. Regul Toxicol Pharmacol 73(2):669–670
4. Lowe NJ (2006) An overview of ultraviolet radiation, sunscreens, and photo-induced dermatoses. Dermatol Clin 24(1):9–17
5. Piccirillo C, Rocha C et al (2014) A hydroxyapatite–Fe_2O_3 based material of natural origin as an active sunscreen filter. J Mater Chem B 2(36):5999–6009
6. Pal A, Hadagalli K et al (2020) Hydroxyapatite—a promising sunscreen filter. J Aust Ceram Soc 56(1):345–351
7. Amin RM, Elfeky SA et al (2016) A new biocompatible nanocomposite as a promising constituent of sunscreens. Mater Sci Eng C 63:46–51
8. Morsy R, Ali SS et al (2017) Development of hydroxyapatite-chitosan gel sunscreen combating clinical multidrug-resistant bacteria. J Mol Struct 1143:251–258
9. Teixeira MAC, Piccirillo C et al (2017) Effect of preparation and processing conditions on UV absorbing properties of hydroxyapatite-Fe_2O_3 sunscreen. Mater Sci Eng C 71:141–149
10. Hadagalli K, Shenoy S et al (2021) Effect of Fe^{3+} substitution on the structural modification and band structure modulated UV absorption of hydroxyapatite. Int J Appl Ceram Technol 18(2):332–344
11. De Araujo TS, De Souza SO et al (2010) Phosphates nanoparticles doped with zinc and manganese for sunscreens. Mater Chem Phys 124(2–3):1071–1076
12. Pyo E, Kim Y et al (2016) A Silver-doped hydroxyapatite for an active sunscreen material. Bull Korean Chem Soc 37(9):1395–1396
13. Abbas N, Manzoor S et al (2021) Investigation of calcium silicate as a natural clay-based sunblock: formulation and characterization. Photodermatol Photoimmunol Photomed 37(1):39–48
14. Egambaram OP, Pillai SK et al (2019) Nanostructured Zn-Ti layered double hydroxides with reduced photocatalytic activity for sunscreen application. J Nanopart Res 21:53
15. Egambaram OP, Pillai SK et al (2023) Structural and photoprotective characteristics of Zn-Ti, Zn-Al, and Mg-Al layered double hydroxides—a comparative study. Cosmetics 10(4):100
16. Ng'etich WK, Martincigh BS (2021) A critical review on layered double hydroxides: their synthesis and application in sunscreen formulations. Appl Clay Sci 208(2021):106095
17. Perioli L, Ambrogi V et al (2005) Anionic clays for sunscreen agent safe use: photoprotection, photostability and prevention of their skin penetration. Eur J Pharm Biopharm 62(2):185–193
18. Perioli L, Nocchetti M et al (2007) Sunscreen immobilization on ZnAl-hydrotalcite for new cosmetic formulations. Microporous Mesoporous Mater 107(1–2):180–189
19. Abdul Aziz INF, Sarijo SH et al (2019) Synthesis and characterization of novel 4-aminobenzoate interleaved with zinc layered hydroxide for potential sunscreen application. J Porous Mater 26(3):717–722

20. Fantini R, Vezzalini G et al (2021) Boosting sunscreen stability: New hybrid materials from UV filters encapsulation. Microporous Mesoporous Mater 328:111478
21. Confalonieri G, Fantini R et al (2022) Structural evidence of sunscreen enhanced stability in UV filter-Zeolite hybrids. Microporous Mesoporous Mater 344(2022):112212
22. Lademann J, Schanzer S et al (2005) Synergy effects between organic and inorganic UV filters in sunscreens. J Biomed Opt 10(1):014008

Chapter 11
Conclusion

It has been conclusively proven that exposure to UVR in sunlight causes skin damage and diseases such as erythema, ageing, pigmentation, and even cancer. Increased awareness of the importance of skin protection from UVR has led to a plethora of sunscreen products in the market, typically containing inorganic and/or organic UV filter actives. Inorganic UV filters have often been favoured over organic UV filters for use in sunscreens due to their broad-spectrum UV protection, high photostability, and good skin compatibility, which is especially suitable for sunscreen users with sensitive and allergy-prone skin. As ZnO and TiO_2 are currently the only two inorganic UV filters approved by regulatory bodies, their use in commercial sunscreens is expected to continue and even increase as organic UV filters lose favour. Current metal oxide-based sunscreens have better aesthetic appeal and formulation stability than earlier generation sunscreens largely due to the reduction of the particle sizes of TiO_2 and ZnO. However, the inherent properties of TiO_2, ZnO, and other semiconducting metal oxides such, as particle size and photocatalytic activity, especially of their nanoforms, have implications for their use in sunscreens related to the photostability and efficacy of sunscreens, human safety and health, and their potential fate and effects in the environment. These concerns have also led to regulation of the use of ZnO and TiO_2 for photoprotection, which dictates their use in sunscreens in terms of their allowed concentrations, particle sizes and crystal phases.

In vitro and in vivo studies reported in the literature have indicated the propensity of TiO_2 and ZnO particles to cause different deleterious effects on plant, animal and human cells and organisms, both without and under UV irradiation. The toxic effects observed were mainly attributed to the interaction of cells with ROS generated on UV irradiation of TiO_2 and ZnO particles and with free metal ions arising on metal oxide particle dissolution in skin, water or soil, especially for Zn^{2+}. However, a review of studies in the literature has largely concluded that TiO_2 and ZnO present a low risk of causing adverse effects to human health, via dermal penetration, and of causing undesirable effects in the environment at the low concentrations that typically occur there. However, it must be noted that human safety and ecotoxicology

N. H. Kera et al., *Inorganic Ultraviolet Filters in Sunscreen Products*,
SpringerBriefs in Materials, https://doi.org/10.1007/978-3-031-64114-5_11

studies did not employ standardized methods for toxicity assessments and were not representative of real-life scenarios for the application, use and release of sunscreens in the environment. Nevertheless, the photoreactivity of the inorganic UV filters used in sunscreen is still a primary concern directly impacting their continued use in sunscreens.

As such, different strategies have been investigated for improving the properties of TiO_2, ZnO and potential alternative metal oxides, such as iron and cerium oxides, for use in sunscreens in order to diminish their undesirable features, such as high photocatalytic activity, while retaining their UV attenuation properties. Coating, surface functionalization, decoration, and metal ion-doping were some methods employed to reduce the photocatalytic activity of metal oxide particles. These methods typically decreased, but did not completely diminish, the photoactivity of metal oxides, through effecting the scavenge of free radicals and ROS or by hindering their formation and movement.

Natural inorganic materials such as hydroxyapatites, hydrotalcites, and calcium silicate have also been investigated as alternatives to traditional metal oxide-based UV filters due to their low photoreactivity, biocompatibility, abundance in nature and relative low cost. Different modification methods such as metal ion doping and incorporation into composites were employed to impart UV attenuation properties to natural inorganic materials. However, natural materials have lower efficacy for UV attenuation than traditional metal oxide-based UV filters, even after modification, and require further improvement before their application as UV filters. The use of alternative metal oxides and inorganic materials in sunscreen is also subject to their approval by regulatory bodies. The complexity of modification techniques and associated costs may also hinder the use of modified materials in sunscreens.

Chapter 12
Future Outlook

The risk associated with exposure to inorganic UV filters, even in the form of NPs, in sunscreens is greatly outweighed by the risks associated with exposure to UVR in sunlight. Subsequently, due to the advantages of inorganic UV filters over organic UV filters, especially as new information comes to light on the undesirable effects of currently approved organic UV filters to human health and the environment, and as no new inorganic UVA/UVB filters have been approved for use by regulatory bodies, TiO_2 and ZnO UV filters are expected to continue to play a prominent role in sunscreen products for photoprotection.

However, there are concerns related to the safety of TiO_2 and ZnO in sunscreens with regard to human health and potential environmental effects that have led to regulation of their use for photoprotection. These concerns are mainly as a result of two inherent properties of metal oxides: the small sizes of their nanoforms and their high photoactivity, especially that of the anatase form. However, there are a number of achievable strategies for circumventing the undesirable features of TiO_2 and ZnO, which provides further impetus for their continued use as UV filters in sunscreens and these are presented below.

There is a need to determine the UV absorption profiles and photoactivities of different forms of ZnO and TiO_2, with regard to properties such as particle size, particle shape, crystal phase and morphology, among others, especially as there are discrepancies in the data reported in the literature. This information will be invaluable for determination of the forms of ZnO and TiO_2 best suited for application as UV filters.

For avoiding the use of NPs in sunscreen products, ZnO and/or TiO_2 microparticles or hierarchical structures with properties, such as shape, size, and thickness, tailored towards their application as UV filters can be used instead. Methods for obtaining these materials, such as through the controlled aggregation of NPs and/or microparticles, should be investigated further.

For reducing the photocatalytic activity of metal oxides, modification of particles by methods such as coating, surface functionalization, decoration and metal

N. H. Kera et al., *Inorganic Ultraviolet Filters in Sunscreen Products*,
SpringerBriefs in Materials, https://doi.org/10.1007/978-3-031-64114-5_12

ion-doping should be further probed, as preliminary studies have indicated these methods to be relatively simple ways of reducing photocatalytic activity. The addition of inorganic and organic materials to formulations that can serve as antioxidants and scavenge ROS generated on UV irradiation of TiO_2 and ZnO is also another method that can be investigated to prevent effects arising due to reactions involving ROS. Inorganic particles, such as silica microparticles, have additional advantages such as increasing the SPF of sunscreen formulations. Natural antioxidants obtained from plants are another option and preferable to use in sunscreens than synthetic antioxidants due to the potential health risks associated with the use of the latter.

Further study of the modification of natural materials is required to enhance their UV attenuation properties. Composites of ZnO and/or TiO_2 and natural materials could be investigated in order to exploit the advantages of both components and enhance the sunscreen formulation properties. The biofunctional and antimicrobial characteristics of natural materials studied as potential UV filters should also be investigated and could offer further advantages for the use of natural materials in cosmetics and sunscreens.

The production and use of sunscreen formulations containing combinations of inorganic and organic UV filters is expected to continue due to the high demand for broad-spectrum, high SPF sunscreens in the market. To this end, the use of inorganic materials such as LDHs and zeolites for encapsulation of organic UV filters for preventing the photodegradation of organic UV filters and for maintaining product efficacy and stability should be further investigated. Further studies of guest–host interactions are required for optimization of the UV attenuation properties and stability of encapsulated organic UV filters. Inorganic materials that can serve as antioxidants can also be investigated for incorporation into combination sunscreens to scavenge photogenerated ROS and prevent ROS-mediated reactions in sunscreen formulations or on the skin.

On a related note, there is an urgent need for standardized testing methods for toxicity assessments used for human safety and ecotoxicology studies in order for scientific and/or regulatory consensus to be reached regarding the safety of using sunscreens containing inorganic UV filters and the potential effects of inorganic UV filters on organisms in the environment. The results of different related studies reported in the literature often vary significantly and comparison is challenging due to differing experimental conditions and setups and due to the lack of correlation between the results obtained from in vitro and in vivo studies. Furthermore, most of the methods employed for toxicology testing of inorganic UV filters do not adequately reproduce real-life scenarios and the results obtained may not be relevant to human safety.